A SAFE AND PROSPEROUS FUTURE

100 years of engineering and geoscience achievements in Saskatchewan

by Edward Willett

A P E G S
Association of Professional Engineers & Geoscientists of Saskatchewan

Association of Professional Engineers and Geoscientists of Saskatchewan
2006

Library and Archives Canada Cataloguing in Publication

Willett, Edward, 1959-
A safe and prosperous future: 100 years of engineering and geoscience achievements in Saskatchewan / Edward Willett.

Includes bibliographical references and index.
ISBN 0-9738758-0-1

1. Engineering—Saskatchewan—History—20th century. 2. Earth sciences—Saskatchewan—History—20th century. 3. Technological innovations—Saskatchewan—History—20th century. I. Association of Professional Engineers & Geoscientists of Saskatchewan II. Title.
TA27.S38W54 2006 620'.0097124 C2005-906935-X

Design by J. Lauder Publishing, Regina, Saskatchewan

Printed and bound in Canada by Houghton Boston, Saskatoon, Saskatchewan

Association of Professional Engineers and Geoscientists of Saskatchewan
2255 - 13th Avenue, Regina, Saskatchewan, Canada S4P 0V6
Tel: 1-800-500-9547 Email: apegs@apegs.sk.ca Website: www.apegs.sk.ca

Achieving a safe and prosperous future through engineering and geoscience

- APEGS mission statement

From the Author

One shouldn't really introduce a book by saying what it isn't, but I think it has to be said:

This isn't a complete and detailed history of engineering and geoscience in Saskatchewan.

Frankly, that would be impossible, even if one wrote a multi-volume set the size of the *Encylopedia Britannica.*

Why? Because engineering is involved in . . . well, everything.

Every building ever built in the province was an engineering project, as was every road, every machine, every bridge, every power plant, every . . . well, you get the idea.

And for its part, geoscience has figured into every oil well and every mine (and in many other projects), each of which comes with its own lengthy history.

This, then, is merely a sampling, a survey of some of the many engineering and geoscience projects undertaken in the effort to build for Saskatchewan a safe and prosperous future. Since the future never gets here, that effort continues!

My thanks to the members of the APEGS Centennial Task Group for their help and encouragement with this project.

Special improved and extended thanks to project manager Sharon Baldwin for keeping everything, including me, on track.

Thanks to Jo Anne Lauder for an outstanding job in putting the words and pictures together.

Thanks to all of the people who graciously allowed me to interview them for this book, and the staff at the various archives and companies who helped locate and allowed us to use the many photos that grace these pages.

Finally, thanks for her support and encouragement to my favourite engineer of all: my wife, Margaret Anne Hodges.

I hope you enjoy reading about some of the successes and challenges from the first 100 years of engineering and geoscience in Saskatchewan—and perhaps come to appreciate more, as I did, just how important are the contributions of these professions to the high quality of life we enjoy.

Edward Willett
May 11, 2006

Foreword

On the occasion of the 75th anniversary of the Association of Professional Engineers and Geoscientists of Saskatchewan (APEGS), I am pleased to have the opportunity to present this work as the Association's gift to the people of Saskatchewan. It is intended to provide a chronicle of some of the professions' most significant contributions to our province during its first 100 years.

As we look back at the development of Saskatchewan into a mature, diverse, inclusive province which has made a significant contribution to Confederation, we realize that the engineering and geoscience professions have worked with the citizens of the province to help achieve this. From the beginning with the construction of the railroads to today's world-class communication systems and booming resource industries, Saskatchewan's engineers and geoscientists have helped make Saskatchewan an excellent place in which to live. Working with their fellow citizens, engineers and geoscientists have contributed to all aspects of Saskatchewan's growth and prosperity.

In this publication we have tried to highlight significant contributions made by the professions. There will, however, be omissions, and to those whose achievements we have not been able to detail, I apologize.

I would like to thank the author, Mr. Edward Willett, the project manager, Ms. Sharon Baldwin, and members of the APEGS Centennial Task Group who have provided guidance and advice during the development of this publication. I would also like to express my appreciation to all those who contributed ideas and material for this effort. Without such participation we could not have achieved the success that is evident as you read *A Safe and Prosperous Future: 100 years of engineering and geoscience achievements in Saskatchewan.*

As president of APEGS during the Association's 75th anniversary year, I congratulate those who have contributed to the development of the engineering and geoscience professions in Saskatchewan and urge all members of the professions to use their professional skills to achieve a safe and prosperous future and to protect the public and the environment.

Jon A. Gillies

APEGS President, 2005-06

Above: Members of the APEGS Centennial Task Group. Left to right, back row: Dena McMartin, Laurence Vigrass, Jon Gillies, Sharon Baldwin (project manager); front row, Robert McDonald (APEGS staff), Myron Herasymuik (task group chair), Ken From.

Notes from the APEGS Centennial Task Group

APEGS consistently encourages the use of the designations "Professional Engineer (P.Eng.)," "Professional Geoscientist (P.Geo.)," "Engineer-in-Training," "Geoscientist-in-Training," "Limited Member (Engineering)," and "Limited Member (Geoscience)" because it is these designations that identify individuals who are properly qualified and licensed to practise and take responsibility for engineering and geoscience work in Saskatchewan. In this publication, however, we have intentionally chosen not to use the designations because, in many cases, the engineering and geoscience activities being described occurred before engineering (1930) and geoscience (1997) registration were available in the province. As well, our objective has been to discuss engineering and geoscience achievements in the broadest sense and these achievements include the work and vision of all members of the engineering and geoscience team–technicians and technologists, professionals from other disciplines, and many others.

The consistent use of units of measurement has been a challenge because this publication covers the century during which metrication took place in Canada. The Task Group initially planned to have all measurements expressed in metric units but it soon became apparent that this was neither feasible nor desirable. The reader will note that most measurements are expressed in metric units, some accompanied by the Imperial equivalent. Some expressions are Imperial only where that seemed to make sense. In determining how measurements should be expressed, the Task Group sought to provide the most meaningful description, keeping in mind that, although many readers of *A Safe and Prosperous Future* will have a scientific background, the book is intended for a general audience.

The Task Group would like to express its thanks to Brian Mlazgar, publications coordinator at the Canadian Plains Research Center, University of Regina, for his assistance and advice at various stages of this project.

TABLE OF CONTENTS

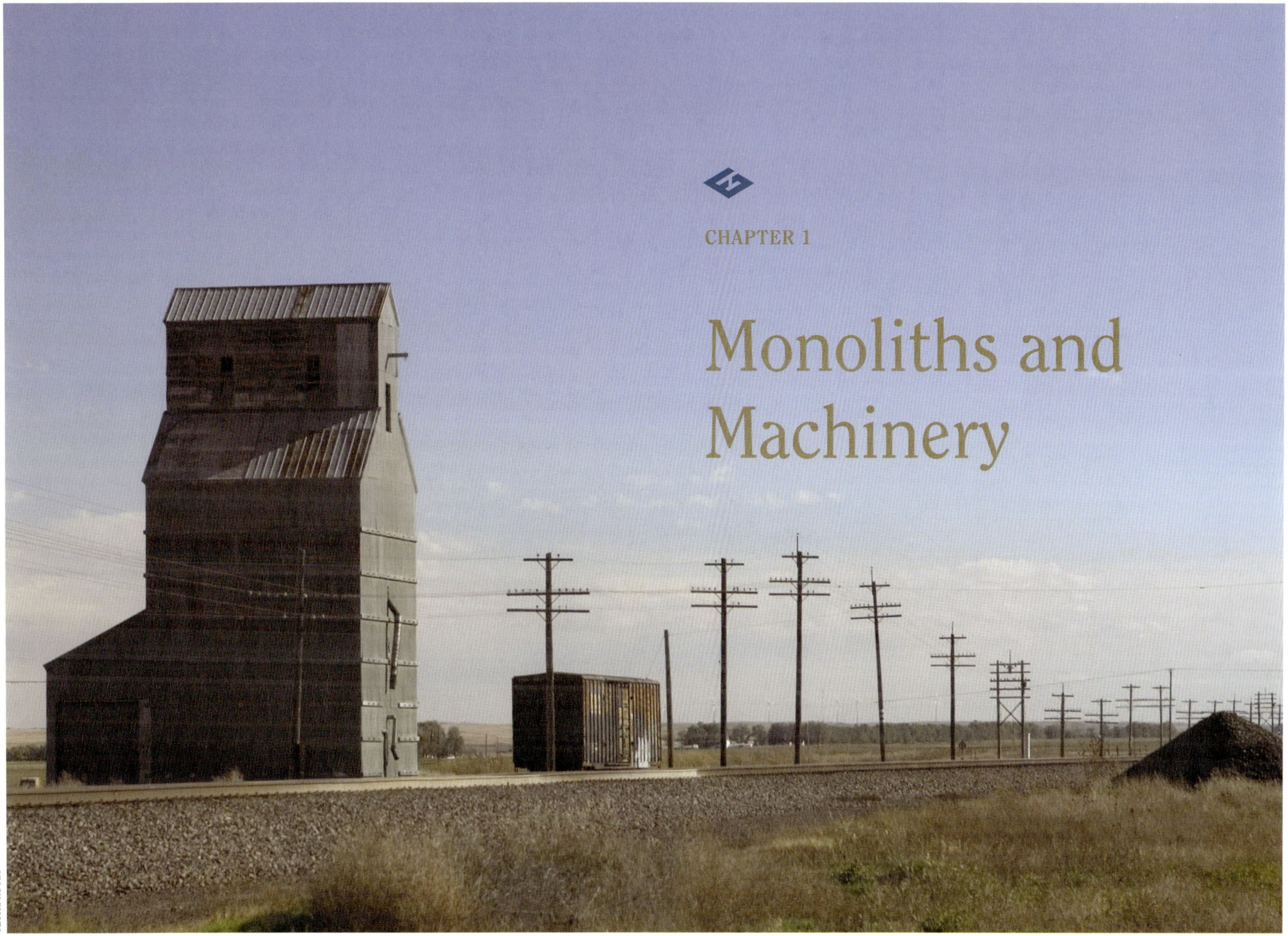

STOCK PHOTO

CHAPTER 1

Monoliths and Machinery

"There is nothing quite so ugly as a grain elevator . . . As a building it is hideous; as a symbol it has the beneficial stars beaten to a frazzle. It is the lightship of the plains."

Building the Grain Elevator Network

Engineers—and engineering projects—were few and far between on the Prairies in the years before Saskatchewan became a province. You didn't need an engineer to build one of the wooden trading posts or military forts that supported the fur trade, and even massive structures like Fort Prince of Wales, which guarded the shore of Hudson Bay at the mouth of the Churchill River, were generally built by men with no formal engineering training.

The first major engineering project in what would become Saskatchewan was undoubtedly the laying of the Canadian Pacific Railway (CPR) transcontinental track, a fairly uncomplicated process on the broad plains. Since most settlers arrived by train after the track was built, few people witnessed its construction.

Train travel changed everything. Thanks to the railway, engineers from eastern Canada could travel to Saskatchewan more easily. And also thanks to the railway, almost every town in Saskatchewan became the site of an engineering project: the construction of a grain elevator.

The good old days

Before the advent of the grain elevator, farmers hauled grain loose or in sacks to flat warehouses, which varied in capacity from 75 to 545 cu. m and were typically 3 to 6 m wide and 6 to 15 m tall. An alleyway passed through the centre of the warehouse, with open bins on either side. The grain was weighed, then dumped into the bins. When it came time to load a rail car, the grain was shovelled out of the bin, moved alongside the car, then shovelled into the car. It often took a whole day of slow, backbreaking manual labour to fill a car.

A grain elevator is simply a vertical rather than horizontal warehouse. The first grain elevator on the Canadian Prairies was the round 909-cu.-m McArthur Elevator in Niverville, Manitoba, built in 1879. The familiar (and easier-to-build) rectangular design appeared first at Gretna, Manitoba, in 1881, courtesy of the Ogilvie Milling Company.

By 1890 the portion of the North-West Territories soon to become Saskatchewan boasted seven elevators. That number swelled with amazing speed (much like the population of the province). By 1900-01 there were 333 elevators in Manitoba and 88 in the North-West Territories. By 1910, there were 707 in Manitoba, 1,004 in Saskatchewan and 285 in Alberta.

The CPR couldn't afford to construct all the country elevators it needed, so it encouraged private companies to build them. The CPR provided free sites for those who erected a standard elevator and promised its cars would not accept grain from anybody but the elevator agent (which had the effect of allowing the grain companies and the railway to maintain a monopoly on grain handling).

STOCK PHOTO

Left page: Elevators sprouted like mushrooms across the Prairie in the early 20th century. Cabri alone had nine of them. *(Saskatchewan Archives Board R-B306.)*

STOCK PHOTO

The location of the elevator usually determined where a town would be built. Anyone travelling Saskatchewan's highways has noticed that there is usually a town—or its remnants—about every 13 km. That's because elevators were built 13 to 16 km apart so that farmers could make one round trip per day to an elevator with a horse and wagon.

The standard elevator

A standard elevator cost $7,000 to $12,000 to build and could be finished in two to three weeks. Standing on a 61-cm-thick concrete foundation, it was typically 9.75 m square and 21 to 24 m high.

To withstand the immense pressure of the wheat (a cubic metre of wheat weighs around 770 kg—the exact density varies with growing conditions), the interior and exterior walls were built of two-by-six boards near the bottom and two-by-four boards near the top, spiked together horizontally. Wood or metal siding covered the exterior. Interior walls were often constructed of two-by-eights in order to carry the load from two adjacent bins. (The fact an elevator can be moved without being dismantled is proof of its structural strength.)

To deliver grain, a farmer drove his wagon (later, truck) up the ramp onto a scale. By weighing the vehicle before and after it unloaded, the elevator agent could determine the weight of the grain and calculate the proper payment.

Air-lifts would raise the front end of the truck, wagon or sleigh, allowing the grain to flow out of the back of the vehicle into a receiving pit. From there it was elevated (hence the name of the structure) by a "leg," a vertical conveyor belt with buckets on it, to the top of the building, then directed by chutes into the appropriate storage bins. There it would stay until it was loaded into a boxcar or grain hopper car.

Power for the leg came first from horses, then from steam engines, and then from gasoline, diesel or kerosene engines or electric motors.

Times change–and so do elevators

Dissatisfied with the service they were getting from the existing grain companies, farmers began forming their own, starting with the Grain Growers' Grain Company (later United Grain Growers) in 1906. But it was when the three Prairie Wheat Pools formed in the 1920s that the number of elevators really soared, peaking in 1933 at 5,758.

A little over a half-century later, in mid-1998, Statistics Canada reported the total had decreased to just 1,052, with 574 of those in Saskatchewan. That number continues to drop.

There have been a number of reasons for the decline. Increased mechanization led to larger farms and a smaller rural population. As hamlets and villages became mere elevator points, branch lines to them became too expensive to maintain. And improved roads—and vehicles—meant the 13-to-16-km spacing was too close.

As the centralization of elevators has proceeded, the elevators themselves have changed.

STOCK PHOTO

Form follows function; aesthetics . . . not so much

Old country elevators have become romanticized in Saskatchewan—they're the most painted and photographed objects on the Prairie. But nobody romanticized them when they were first built.

Early writers used terms like "helpful, if unpicturesque." Walter Herbert, an early publicist for the Wheat Pools, described elevators as "stark and rather unbeautiful," "drab and ugly," and "architecturally . . . nothing to inspire delight." (Dommasch, p. 7.)

H.F. Gadsby probably summed up the populace's early opinion of grain elevators best in an article in the *Regina Morning Leader* on April 25, 1912. Gadsby wrote:

"Although the common crib elevator is not as beautiful as the parliament buildings at Regina, it is perhaps as important a factor in the life of the Saskatchewan people. There is nothing quite so ugly as a grain elevator, a big galvanized iron box, surmounted by a smaller box with a tin roof, the whole painted a sullen weather-beaten red . . . As a building it is hideous; as a symbol it has the beneficial stars beaten to a frazzle. It is jammed with fatness and reserves of satisfaction like Joseph's granaries. It usually holds thirty or forty thousand bushels of grain and to that extent it represents, happiness, hope and the reward of labor . . ." (Saskatchewan's Top News Stories: Politics.)

Although, "You'd be surprised at how little the concept has changed," said Charles Dietrich, manager of engineering for the Saskatchewan Wheat Pool. "Probably the biggest change has been their physical size, and the material used to construct them."

Grain companies have experimented with different materials for many years. A wooden grain elevator has a life expectancy of about 40 years, the time needed to run through 150 000 to 180 000 cu. m of grain. Despite its immensely strong construction, it takes a pounding from being filled and emptied.

The Saskatchewan Wheat Pool constructed two all-steel elevators in the 1960s, one at Kenaston and one at the flour mill in Saskatoon. A third was built in Bruno in the early 1990s. However, the cost proved to be substantially higher than for a standard wooden elevator.

The Wheat Pool began experimenting with pre-cast concrete slabs in 1968. The prototype had an increased storage capacity, and could fill a boxcar 10 times as fast as a standard elevator. It also needed little maintenance.

Today's elevators, many of which have been built within the last decade or so, are behemoths made of slip-form concrete and shaped like silos (harking back to the round shape of the first Prairie elevator in Niverville). Concrete, Dietrich said, became the material of choice because of its durability and resistance to fire.

The new elevators have 10 times or more the capacity of an old wooden elevator, and can load cars faster, too. Whereas an old elevator could fill a boxcar with 80 cu. m in 15 to 20 minutes, a modern elevator can disgorge more than 1000 cu. m of wheat in an hour and spot 50 to 100 hopper cars at a time. The new elevators also typically include drying and cleaning equipment, enabling them to dry up to 50 tonnes of grain per hour and clean 120 tonnes at the same time.

"The other noticeable thing," Dietrich said, "is we are able to load a unit train, a maximum car-length train that goes directly to the coast without stopping off at a number of facilities."

There's one other major technological advantage a modern elevator holds over a traditional one. "Automation and remote controls make grain handling today more efficient by reducing the labour needed to handle a unit of product," Dietrich said. "The plant is entirely remotely controlled . . . with electronic controls and computers. It gives us some side benefits in terms of the maintenance monitoring so we can tell how long a piece of equipment was running. We can even trace the operator's steps in the event that a failure occurred."

In smaller Saskatchewan communities in particular, grain elevators continue to be the largest, most visible, and most expensive examples of the work of engineers on the Prairies.

But not far away from the elevator in most towns, you'll find a farm machinery dealership . . . and more examples of how engineering has benefited Saskatchewan agriculture.

Saskatchewan's largest elevator

In the early 1970s a group of about 1,450 farmer/shareholders in the Weyburn area saw a need to improve the existing grain handling and transportation system—so they built their own elevator, the largest in Saskatchewan.

The Weyburn Inland Terminal began operation in 1976. It's an 11-storey, 57-m-tall giant, whose total capacity now (with the addition of several storage bins to its original 36 370-cu.-m structure) is 112 000 tonnes, or around 150 930 cu. m, making it even larger than one of the Vancouver port terminals. Two of its legs can elevate 800 tonnes per hour. Two others run half that fast. In one hour it can dry 60 tonnes of grain and clean 200 tonnes. It can also load 600 tonnes per hour, or 100 hopper cars per day.

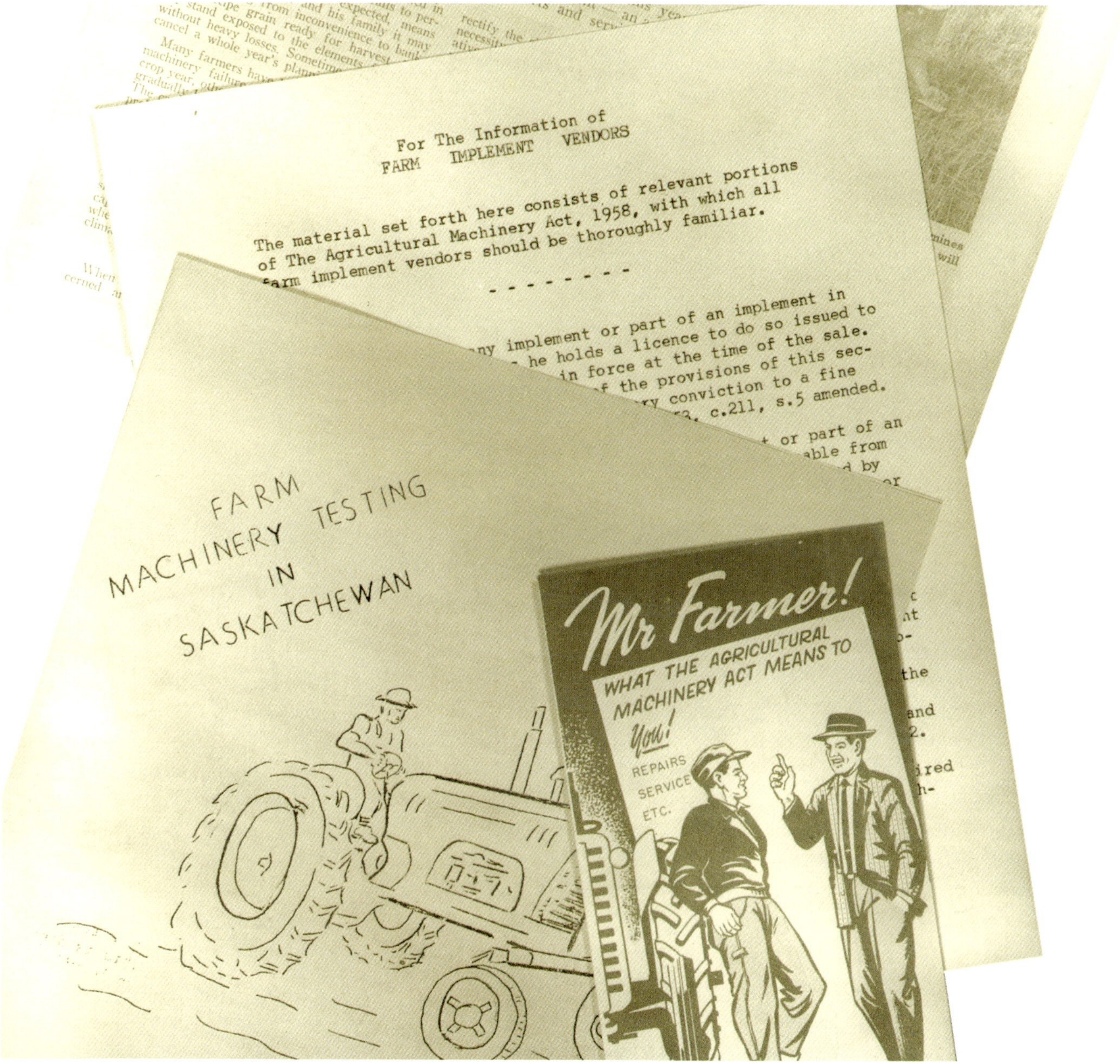

Inventors and Innovators

Engineering Better Farm Equipment

Farming in Saskatchewan is highly mechanized. Historically, much of that machinery has been made elsewhere—and elsewhere often has very different conditions than these dry, dusty plains.

If necessity is the mother of invention, she's spent a lot of time in Saskatchewan. Faced with recalcitrant equipment day after day, many a Saskatchewan farmer has taken up whatever tools and equipment he had at hand and tried to come up with a better solution. Some of those inventions were used only by the farmers who invented them, but more than a few have given birth to manufacturing companies large and small.

Left: Pamphlets and other promotional material put out by the Saskatchewan Agricultural Machinery Administration in the late 1950s. *(Saskatchewan Archives Board R-B4719.)*

Above: A Saskatchewan farm boy shows off an early gasoline tractor and plow. *(Saskatchewan Archives Board R-A107-2.)*

Right page: An early steam traction engine pulls a plow across a Saskatchewan field. *(Saskatchewan Archives Board R-A108.)*

The requirements of the district

By the turn of the 20th century, blacksmiths, small iron works and foundries were producing a limited range of simple farming equipment: plows, land packers, harrows, water pumps, etc. The *Grain Growers' Guide* noted in 1920 that "in most cases their output is determined by the requirements of the district they serve and in many cases they manufacture machinery which is patented by a local farmer for use in the immediate district." (Wetherell and Corbet, pp. 44-45.)

The "requirements of the district" drove much of the farmer innovation in Saskatchewan. Consider tillage.

By the end of the First World War, dryland Prairie farmers had found that the equipment manufactured by the "full-line firms" (equipment manufacturers that carried farm equipment of all types, as opposed to the "short-line firms" that carried only a limited selection of more specialized equipment) didn't meet their needs.

As Donald Wetherell and Elise Corbet put it in their book *Breaking New Ground*, "As a group, farmers showed a strong inventive and mechanical bent, and for a great number of them, the natural joy in the exercise of imagination found expression in working with machinery. Since repair parts were expensive or sometimes difficult to obtain, many were forced to adapt materials at hand or devise home-made solutions to keep equipment running. These farmer inventions and adaptations were widely recognized as holding important clues to improved agricultural technology." (Wetherell and Corbet, p. 48.)

The professional scientists and engineers at experimental farms such as the one in Swift Current

helped these inventive farmers by testing and improving the homemade machinery that came their way. The Agricultural Engineering Committee (AEC), created by the same act of Parliament that established the Prairie Farm Rehabilitation Administration in 1935, noted in 1936 that "the majority of progressive ideas for machinery improvements has come from farmers." The AEC also noted that there were more than 200,000 farmers on the Prairie, "at least one per cent of whom will have one or more useful and progressive ideas." (Wetherell and Corbet, p. 48.)

The committee's estimate may have been low. Between 1905 and 1980, Saskatchewan residents patented 3,200 inventions. Almost a third of those patents were awarded in the 1920s, and most were related to agricultural equipment. (Most were for improvements on existing equipment rather than designs for new equipment, possibly because the general principles of a new design are often unpatentable.)

Some ideas were never patented because the farmers lacked the training and skill to engineer their inventions into the form required for patenting and production. One such unpatented invention was the swather, which revolutionized grain harvesting on the Prairie.

Harvest difficulties

In the early years of the 20th century, harvesting involved three labour-intensive steps: reaping/binding, stooking and threshing. The reaper/binder (known as a binder in Western Canada) cut the grain and bound it with twine into sheaves. So that the sheaves could dry, they were set up by hand (stooked) in the field. Finally, the dried sheaves were gathered and threshed.

Prairie inventors formed numerous companies to market various mechanical stookers. None succeeded. They had more luck with automated sheaf loaders, which reduced the number of hired men and horses needed to pick up the sheaves and take them to the threshing site.

The threshing machine also underwent constant improvement.

Above: Visitors to an exhibition (possibly Saskatoon, ca. 1925) stroll past the display of agricultural machinery. *(Saskatchewan Archives Board R-A5705-2.)*

Right page: Many Saskatchewan farmers have improved or even invented the machinery they use on their farms—like this 1950 homemade stacker from the Maple Creek district. *(Saskatchewan Archives Board R-A11220-4.)*

When Saskatchewan became a province, huge threshing machines were driven by belts powered by giant steam traction engines. In 1908 the new province had about 3,200 of these monsters, which were so expensive to buy and operate that farmers formed partnerships to purchase them, or else hired one of the custom threshing crews that moved continually across the Prairies like modern-day nomads.

But the threshing machine's fate was already sealed. Combines, machines that combined harvesting and threshing, were in use in California by 1890. Those early machines weren't suitable for the Prairies because they were designed to harvest a crop standing in the field, which meant it had to be evenly ripe—and few fields ripened evenly over Saskatchewan's short growing season. Heavy, unripe grain, often found in low spots in the rolling fields of the parkland, or green weeds, common everywhere, could jam the combine.

But a self-propelled, gasoline-powered Massey-Harris combine tested at the Swift Current Experimental Farm in 1922 proved so successful that the farm purchased one for itself. Massey-Harris, which hadn't marketed the combine on the Prairies before that, soon sold more. By 1928, almost 2,400 combines were in use in Saskatchewan.

Even so, many farmers felt the newfangled device was only suited to dry areas where the crop could ripen fully on the stalk. What finally changed their minds was probably the most significant farmer invention ever to come out of Saskatchewan: the Hanson swather.

The amazing Hanson brothers

In 1908 North Dakotans Ole and August Hovland received a patent for a swather and a "travelling thresher." In 1909 they formed a company and built prototypes of the two machines, which operated separately.

The Hovlands' swather had a 4.5-m-wide section with a cutting bar on either side of the front of a tractor. Canvas belts running on a table carried the cut grain to the centre of the machine, laying it on the stubble in a row that emerged between the tractor's rear wheels. (The Hovlands had noticed that cut grain lying loose on top of stubble dried quickly.)

The travelling thresher, which followed behind, picked up the grain and threshed it on the spot. (The pick-up wheel was ground-driven, but a motor powered the threshing operation.)

The Hovland designs worked, but garnered little interest: the swathing/threshing model was too foreign to what farmers were used to in 1909. By the 1920s, though, with combines widespread and the days of stooking and giant threshing machines quickly fading, farmers were more open to new ideas. Enter Ellert and Helmer Hanson, who farmed at Lajord.

Having grown up close to where the Hovlands had carried out their experiments, the Hansons knew about the Hovlands' swather/thresher combination. Although lacking formal education, they both had plenty of mechanical skill and experience: Helmer became the family blacksmith at age 15 and Ellert held an engineer's certificate to operate boilers and engines. They also had a strong interest in modernizing farming methods.

Ellert, convinced tractors were the way of the future, sold his horses by 1919. By 1921 he had made a 30-run seed drill out of two 20-run drills because he wanted an implement large enough to make good use of his tractor. That same year Helmer made a 6.4-m disc harrow, later adapted for commercial use by International Harvester. Together the brothers produced many other attachments and improvements.

In 1925, spurred by the difficulty of getting good harvest help, they started thinking about the "possibility of picking up a swath and threshing it with a standard combine redesigned in part to handle such a row," Helmer later said. (Wetherell and Corbet, p. 167.)

They began by creating a swather. A cutting bar and reel knocked down the grain; a canvas belt running over a table carried the cut grain to the centre and laid it on the stubble in a windrow.

An American Abell plowing and threshing engine, photographed in Toronto, which was "popular for plowing in the Regina gumbo for breaking as front wheels are close together and packs sod so mouldboard on breaker plows clean." *(Saskatchewan Archives Board R-B7811.)*

In 1928 an engineer from International Harvester Company (IHC) came to see the Hansons' swather at their farm. "He looked smart and he was smart," Helmer recalled. (Wetherell and Corbet, p. 167.) The engineer asked one basic question: what would they do differently if they were to build another swather? The Hansons told him they would make the rear opening larger, to make it easier for "heavy, tangled or weedy" material to clear.

Shortly thereafter, IHC began selling its first swather, based on the Hanson design—but with a larger rear opening. When IHC came into the picture, Helmer Hanson said, "we felt that two years of strenuous effort and awful chances taken with our crops had paid off." (Wetherell and Corbet, p. 168.)

The Hansons now had to adapt the combine. They first modified the machine so it would pick up sheaves, because "that would cut the cost of threshing a great deal, but it was a compromise method if our row (swath) failed us." (Wetherell and Corbet, p. 168.)

Applying a sheaf pickup to a combine was new; J.I. Case tooled a special combine cylinder to the Hansons' specifications. It worked. They used it for two seasons, then changed it to pick up a straight swath, having become convinced that a straight swath would keep just fine, making sheaves unnecessary.

By the 1928 harvest season, there were 4,400 combines on the Prairies, almost half with pickups, and almost 2,000 swathers. The Depression put a crimp in the sale of new equipment, but by the late 1940s swathers were in general use.

Not much formal training

There were many farmer-inventors, but engineering knowledge was in short supply on the Prairies in the early years. For all their success, the Hansons were aware of what they didn't know. "All our blueprints for the whole thing were stored between our ears, no drawings, each of us could see in his mind the complete machine and swather ready to go to work," Helmer said. "Not being engineers or draftsmen, we did it the only way left to an untrained mind . . . We were handicapped by limited mathematics, true, we worked out our own problems but we probably did it the hard way—sometimes by sheer reasoning." (Wetherell and Corbet, p. 50.)

The Department of Agricultural Engineering at the University of Saskatchewan wasn't established until 1909. The university provided training both through degree programs and short courses for farmers, but even by 1939, Evan Hardy, an agricultural engineer at the university, felt that a lack of "competent engineers" was a barrier to the development of farm equipment manufacturing on the Prairies. (Wetherell and Corbet, p. 50.)

Even farmers with some engineering knowledge were hampered by the conditions under which they had to work. Most farms, after all, had few tools, no welders, no acetylene torches, and no electricity until after the Second World War. Would-be equipment manufacturers on the Prairies faced similar shortages of power

Why no patent?

Why didn't the Hansons apply for a patent? Said Helmer,"A specific patent is not of much value unless you fabricate and sell and get returns that way. You may be sure that someone will snipe you from any angle possible. You may have to go to court to defend yourself. With this in mind we welcomed engineers from any of the implement companies, they were free to copy any or everything." Twenty years after IHC did just that, Hanson noted, "We still have our farms. We did not lose them in court trying to defend specific patents. It is easy to overestimate an idea that is so slow in coming that it looks too big when it does come." (Wetherell and Corbet, p. 49.)

"All our blueprints for the whole thing were stored between our ears."

and raw materials, which in turn limited the possibility that the inventions of the farmers would ever find widespread use. Most manufacturing was therefore on a small scale for local use.

Things look up

After the Second World War several short-line companies found success. Many could trace their roots to the earlier period when equipment was manufactured locally by a blacksmith.

Take Schulte Industries. John Schulte took over his father's blacksmith business in Englefeld in 1938. In 1942 he began manufacturing brush-cutting and -clearing equipment, and in 1948 he developed and tested a front-mounted snow blower, which sold increasingly well through the 1950s. Then in 1963, Schulte designed and patented the Schulte conveyor-type rock-picker, the mainstay of the company for years to come.

Today the business is known as the Schulte Group of Companies, and offers a diverse line of equipment for farm, industrial and commercial work, sold throughout North America and around the world . . . but it's still headquartered in Englefeld, Saskatchewan, where John Schulte was a blacksmith.

Another example was Leon Malinowski, who took over his father's blacksmith shop in Bankend and began manufacturing stock troughs, fuel tank stands and, by 1952, bulldozer blades of his own design that featured a quick-mounting system and special features to prevent breakage if the blade hit a heavy object.

Malinowski incorporated Leon's Manufacturing in 1965 and moved to Yorkton in 1967. Today Leon's Manufacturing boasts 9290 sq. m of production space and products ranging from 'dozer blades to land scrapers to manure spreaders.

Rod weeders and land packers

George Henry Morris wasn't a blacksmith, but he had the same background of small-town ingenuity. Born in Bangor in 1904, he farmed with his father until 1924, studied motor vehicle mechanics in Winnipeg, then returned to Bangor to open a garage and implement dealership with his brother.

His customers kept complaining to him of problems with rod weeders. A rod weeder uses a subsurface rotating rod to tear out tap-rooted weeds such as Canada thistle. Because it rotates in a direction counter to a forward-advancing wheel, the rod weeder flips the weeds onto the surface. It also provides some soil packing and doesn't pulverize the soil, all things that made it a popular Saskatchewan implement in the 1920s—except the rods kept breaking, bending or getting thrown out of the soil when they encountered one of the prairie's plentiful stones.

Morris invented an automatic spring-loaded device that permitted a section of the rod weeder to trip backwards when it hit a stone or root, then return automatically to its position in the soil. By 1929 he had patented the device. When the full-line companies turned him down (stones just weren't a problem in most places), he began producing his own 2.4-m horse-drawn rod weeders.

Morris contracted the building of the Morris Rod Weeder to custom manufacturing shops for 30 years, then built his own plant in Yorkon in 1959. Today, Morris Industries Ltd., a leading manufacturer of air seeder, air drill and tillage technology, markets its products worldwide.

A trainload of traction engines and threshers makes its way to Swift Current. *(Saskatchewan Archives Board R-B9691.)*

Above (left): Machinery at Friggstad Manufacturing Ltd. in Frontier, Sask., in September 1980. *(Saskatchewan Archives Board R-PS80-1528-13.)*

Above (right): A Rock-O-Matic rock picker, ca. 1969. *(Saskatchewan Archives Board R-PS69-699-103.)*

There are many such stories in Saskatchewan. Emerson Summach trained as a machinist and then farmed at Asquith. In 1947 he developed the idea for a coil land packer, a heavy, continuous coil of steel pulled behind a tractor. (Packing soil helps moisture move into the seedbed and provides a better environment for the emerging roots.) The full-line companies told him a coil packer would be too expensive to manufacture, so in 1951 with the help of his brother Kenneth, Emerson launched his own company. Using materials taken mostly from a junkyard, they built their own coil-rolling machine. In 1952 they manufactured 67 packers in a relative's machine shop—and sold them all. Today their company, Flexi-Coil, which was purchased by CNH (Case New Holland) in 2002, makes a wide range of seeding, tillage and chemical application equipment.

Some of Flexi-Coil's equipment sprang from the inventive mind of Olaf Friggstad, a Frontier-area farmer. In 1970 Friggstad started Friggstad Manufacturing Ltd. with his two sons. Drawing on their years of experience, they invented a cultivator designed to cope with stony land and trash cover. Flexibility between the sections ensured even penetration of the soil. The firm diversified into rock-pickers, bale wagons, land levellers, tine harrows and air seeders. It went into receivership in the 1980s . . . but its assets, including its machinery designs, were purchased by Flexi-Coil.

A revolutionary air seeder

Another industry leader that grew out of one man's creativity and experience is Bourgault Industries. In 1969 Frank Bourgault, then 44, was a farmer and a partner in the

Massey Ferguson farm implement dealership in St. Brieux. The St. Brieux area is extremely stony, and Frank, who had a reputation as an excellent farm equipment mechanic, was familiar with all of the problems cultivators faced under those conditions, including excessive shank bending, poor trash clearance, lack of penetration and poor weed kill. He felt he could build a better cultivator than what was then available.

He built a prototype in 1970 for just $1,300 and tested and developed it over the next few years. It was so well received that in 1973 Bourgault, his eldest son, Gerry (then in his final year of mechanical engineering at the University of Saskatchewan), and his business partner, Don Coquet, incorporated F.P. Bourgault Industries Ltd. and began constructing a manufacturing facility. The business began with a work force of 10, mostly welders and metal fabricators who were also area farmers.

Gerry (who had designed a mounted tine harrow arm to complement the cultivator as his fourth-year design project) joined the company full time in May of 1975. That same spring, the first major run of 75 Bourgault cultivators and around 2,500 mounted harrows went into service in Western Canada.

Another University of Saskatchewan engineering graduate, Richard Coquet, joined the company in the spring of 1976. Gerry Bourgault and Richard Coquet, along with Frank Bourgault and Don Coquet, next developed the industry's most advanced dual-purpose cultivator, the Bourgault Vibra-Master. From 1976 to 1979, the company manufactured and sold almost 1,000 cultivators and nearly 9,000 harrow units, and employment swelled to 75 people.

In 1978 Frank Bourgault acquired the rights to a unique seed metering system invented by Jerome Bechard, also from Saskatchewan. This became the heart of the Bourgault Air Seeder. The first seven production "Model 138s" were manufactured in the fall of 1980. The Bourgault Model 138 was the first air seeder to be towed behind the cultivator, which gave the operator an unobstructed view of all cultivator shanks. As well, it rode on three large flotation tires that applied no more pressure to the field than someone walking. It could also be easily disconnected.

Further enhancements in cultivators and air seeders followed. Today Bourgault Industries, with Gerry Bourgault as president, remains a world leader in one-pass air-seeding technology. The company owns and operates assembly and distribution facilities in Canada, the US and Australia, and employs more than 300 people.

Equipment testing

While farmers have come up with many of the major innovations in agricultural machinery, engineers are almost always involved at some point in helping to test, improve or develop the machinery for production.

Findings from the equipment tests carried out at the Swift Current Experimental Farm in the 1920s were published in the station's annual reports or in special publications. The first combines were tested there in the 1920s. Similar tests were carried out on other locally developed machinery. Universities also tested and developed equipment. The Agricultural Engineering Committee, mentioned earlier, included agricultural engineers from both the government and universities.

Above (left): A Leon front-end loader displayed at the 1982 Western Canada Farm Progress Show in Regina. *(Saskatchewan Archives Board R-PS82-0818-R4-36.)*

Above (right): A Schulte stone picker on display at the 1982 Western Canada Farm Progress Show in Regina. *(Saskatchewan Archives Board R-PS82-0818-R4-33.)*

In 1949 the Saskatchewan legislature passed an act that included provisions for the appointment of an Implement Engineering Committee to test equipment and parts for their suitability for use in Saskatchewan. It wasn't until 1958, however, that the Agricultural Machinery Administration (AMA) was set up to administer the legislation and test equipment in the laboratory and under field conditions. Any problems identified were reported to the manufacturers.

AMA, the first government testing agency in Canada, regularly issued pamphlets that included test results, recommendations and the response of the manufacturers. By the early 1960s, about 15,000 copies of each pamphlet were being distributed in Saskatchewan.

Farm manufacturers weren't wildly enthusiastic about the AMA. Neither were the Liberals, who thought the government had no business spending money on testing farm machinery. When the Liberals formed the government after the 1964 election, they abolished AMA's testing activities and transferred the testing equipment and some staff to the University of Saskatchewan.

But the Barber Royal Commission concluded in 1967 that the AMA had been valuable and should be reinstated, and when the New Democratic Party was re-elected in 1971, it re-established a farm equipment testing facility, now under the name Prairie Agricultural Machinery Institute (PAMI).

Set up in 1974, PAMI was based in Humboldt, with a research branch at Lethbridge and, after 1977, at Portage la Prairie (all three Prairie governments shared the cost). Government funding dropped to about 30 per cent of PAMI's budget in 1985, with the remainder coming from fee-for-service work. That same year PAMI discontinued the grant-funded machine evaluation program, focusing instead on applied research and machinery development. PAMI's current staff of 45 includes eight engineers: three mechanical engineers and five agricultural and bioresource engineers.

PAMI carries out approximately 100 projects each year. Applied research projects are usually done for producer groups to help them capitalize on opportunities or overcome challenges. For example, two recent projects focused on livestock watering from surface sources and swine manure injection technology.

. . . a farmer's good idea, properly engineered, becomes the basis for a thriving company.

Machinery development projects usually focus on helping manufacturers produce more efficient and effective machinery, and—in time-honoured tradition—to help adapt machinery developed elsewhere to suit conditions in Saskatchewan and Manitoba.

Top Gun

A recent example of machinery developed with the assistance of PAMI's engineers is the Highline Top Gun, which helps control odour around hog, dairy and sewage lagoons by covering them with a 25-cm-thick layer of straw, processed from a round bale. The Top Gun can process one round bale every five minutes, fast enough to go through 80 to 100 round bales—and cover a lagoon 53 m by 182 m—in a single day.

Like many of the companies already mentioned, Highline sprang from the inventive mind of a single man—in this case, Vonda-area farmer Rosaire Bussiere, who built his first rock-picker in 1961 and set up a company called Rock-O-Matic to manufacture it.

In 1976 the company developed the first combined rock-picker and windrower in North America. Two years later Smith-Roles bought Rock-O-Matic. Bussiere's son Ray continued to manage the Vonda plant, however, and when Smith-Roles went into receivership in the late 1980s, Ray Bussiere bought the Vonda plant and renamed it Highline Manufacturing Inc. He also bought an established Manitoba manufacturer of round bale equipment and packer/harrow bars called Laurier Manufacturing in order to expand the line beyond rock-pickers.

Today Highline sells two models of rock-pickers, three models of bale processors, a full line of twin-auger vertical feed mixers, a bale mover and a large selection of belt conveyors—and continues to research and develop new machinery like the Top Gun.

It's a fitting example to end this chapter with, because it's a classic Saskatchewan tale: a farmer's good idea, properly engineered, becomes the basis for a thriving company that continues to develop and market innovative agricultural equipment for the Prairies—and the world.

It may be a bit of a cliché—but it's a good one.

CHAPTER 1 SOURCES

Books

Dommasch, Hans, Ed. *Prairie Giants*. Saskatoon, 1986. Western Producer Prairie Books.

Driver, Deana. *Just a Bunch of Farmers*. Weyburn, 2001. Weyburn Inland Terminal Ltd.

Korol, Todd (photography) and Butala, Sharon (essay). *Harvest*. Saskatoon, 1992. Fifth House.

Lamont, Cecil. *Prairie Sentinels*. Winnipeg, ca. 1941. Northwest Line Elevators Association.

McDonnell, Greg. *Wheat Kings: Vanishing Landmarks of the Canadian Prairies*. Erin, ON, 1998. Boston Mills Press.

Strange, H.G.L. *A Short History of Prairie Agriculture*. Winnipeg, 1954. Searle Grain Company Ltd.

Wetherell, Donald and Corbet, Elise. *Breaking New Ground: a Century of Farm Equipment Manufacturing on the Canadian Prairies*. Saskatoon, 1993. Fifth House.

Websites

Bourgault Industries. (April 4, 2006.) <http://www.bourgault.com)

Civilization.ca – Canada Hall – Saskatchewan Grain Elevator. *Civilization.ca – Web site of the Canadian Museum of Civilization and Canadian War Museum*. (April 4, 2006.) <http://www.civilization.ca/hist/phase2/mod2e.html>

Flexi-Coil. (April 4, 2006.) <http://www.flexicoil.com/>

Highline. (April 4, 2006.) <http://www.highlinemfg.com/>

Leon's Manufacturing. (April 4, 2006.) <http://www.leonsmfg.com/>

Morris Industries. (April 4, 2006.) <http://www.morris-industries.com>

Prairie Agricultural Machinery Institute. (April 4, 2006.) <http://www.pami.ca/index.htm>

Saskatchewan Agricultural Hall of Fame. (April 4, 2006.) < http://www.sahf.ca/>

Saskatchewan Elevators. *University of Saskatchewan Archives*. (April 4, 2006.) <http://scaa.usask.ca/gallery/elevators/>

Saskatchewan's Top News Stories: Politics. *Saskatchewan News Index 1884-2000*. (April 4, 2006.) University of Saskatchewan Libraries. <http://library.usask.ca/sni/index.html>

Saskatchewan Wheat Pool. (April 4, 2006.) <http://www.swp.com/>

Schulte Industries. (April 4, 2006.) <http://www.schulte.ca/>

Personal interviews conducted by the author

Dietrich, Charles. Manager of Engineering, Saskatchewan Wheat Pool. November 9, 2005.

CHAPTER 2

Better by a Dam Site

An earthmover rumbles past the Gardiner Dam control shaft superstructures in 1964. *(Saskatchewan Archives Board R-B4476-2.)*

STOCK PHOTO

> “Any project that keeps my tender bottom off the seat of an outdoor privy at forty below is worth every cent no matter what the cost.”

In 1847 Captain John Palliser, a rich Irishman, came to the western plains of North America to see what the new land might have to offer (and to shoot bison). Upon his return to England, he wrote a popular book entitled *Solitary Rambles and Adventures of a Hunter in the Prairies.*

Palliser so enjoyed himself in the West that he decided to return—preferably at someone else's expense. He presented an exploration plan to the London Royal Geographical Society, which in turn persuaded the Colonies Ministry to assemble a scientific team to accompany Palliser on an expedition more or less along the 49th parallel of latitude.

Over three years beginning in 1857, Palliser explored the mountains, plains and rivers of what would eventually become Western Canada. Among other things, he identified “Palliser's Triangle,” a very dry region of the prairie stretching from the Rockies to southern Manitoba. He thought this near desert, too dry to support normal agriculture, might hinder western expansion.

In 1858 a second expedition set out from Upper Canada, led by Henry Youle Hind, an Englishman who had immigrated to Toronto in 1846. With a mandate from the Geological Survey of Canada similar to that of Palliser, he and an engineer, Simon James Dawson, made observations in the Assiniboine, Souris, Qu'Appelle and South Saskatchewan river valleys.

Both Hind and Palliser noted that the Qu'Appelle River started in a valley just 18 km from an elbow of the South Saskatchewan River, and both had the same idea: if you could build a dam to divert water from the South Saskatchewan River to the Qu'Appelle River (not an easy task, since the source of the Qu'Appelle was 26 m higher than the typical summer level of the South Saskatchewan), you could create a navigable channel all the way from the Rocky Mountains to lower Lake Winnipeg.

Thirsty settlers

With the coming of the railway, the need for long-distance waterways diminished, and so did one rationale for building a South Saskatchewan River dam.

But the need for water for other uses increased with the population. More than 700,000 people came to Saskatchewan and Alberta between 1901 and 1911. Regina swelled from 2,300 to 30,000 people, and Moose Jaw from 1,500 to 14,000. All those new citizens needed water.

So did all the new farmers, who were discovering that Palliser's Triangle was as dry as he'd said it was. Droughts occurred regularly.

During its 1919-20 session, the Saskatchewan legislature passed an act “respecting a supply of water from the South Saskatchewan River.” A commission proposed a $5.5 million pipeline. Voters in Regina and Moose Jaw liked the idea, but those in smaller communities didn't. The project never got off the ground.

But then came the worst drought of all. Between 1929 and 1937, many bodies of water on the Prairies dried up altogether; others reached record lows.

STOCK PHOTO

For example, before the drought, Crane Lake in southwestern Saskatchewan was 9.6 km long and 4.8 km wide, and the headquarters of 76 Ranch, one of the largest in Western Canada. It went bone dry. Cypress Lake, 9.6 km long and 3.2 km wide, shrank to a muddy water hole. Swift Current Creek, the Qu'Appelle River and the Souris River turned into a series of stagnant pools over much of their length. Shallow wells went dry. Water had to be hauled long distances. (Spence, p. 74.)

In response to the suffering caused by the drought, in 1935 the federal government passed the *Prairie Farm Rehabilitation Act*, creating the Prairie Farm Rehabilitation Administration (PFRA) to deal with the water shortage problem and to put the region's available water supplies to the best economic use.

A dam here, a dugout there

Most early PFRA projects were farmer-initiated. A farmer would locate a site for a dugout or dam, then apply for funding. A qualified PFRA officer would inspect and test the site, then the successful applicant would construct the project in accordance with engineering specifications—and the finished project would receive a final PFRA inspection to be sure it was properly built.

Larger projects usually focused on irrigation. Among the first in Saskatchewan was a dam across the Frenchman River near the village of Val Marie. The 1939 Annual Report of the PFRA describes it: "This project provides for the irrigation of 6,000 acres of river-bottom flats . . . by means of a 3,600-foot dam across the Frenchman River, constructed in 1936, which stores some 6,000 acre-feet of water." A later dam constructed just above that reservoir provided irrigation water for additional lands at a higher elevation. (Spence, p. 86.)

STOCK PHOTO

Another dam built on the Frenchman River that same year, 80 km west of Val Marie near Eastend, provided water to irrigate 1295 ha. The Cypress Lake reservoir, with a capacity today of 128.652 million cu. m, was constructed later to provide water to supplement the Val Marie and Eastend reservoirs, for other irrigation projects on the Frenchman River and Battle Creek, and to maintain a steady stream flow for watering stock. (Spence, p. 86.)

In 1937 amendments to the *Prairie Farm Rehabilitation Act* extended its scope. New activities included moving settlers from sub-marginal or non-arable land to new locations and turning the lands they left into community pastures.

In the spring of 1938 George Spence, a veteran of 20 years of federal and provincial politics, was named director of PFRA. He would hold the post for nine years. Spence had a passionate interest in water conservation that dated back to his own experiences as a homesteader.

Construction of new irrigation projects continued early in Spence's term. In 1938, Maple Creek irrigation project's three reservoirs and many kilometres of canals put to use water that had previously drained from the north slope of the Cypress Hills into Big Stick Lake and evaporated.

A 20-m-tall earthfill dam on Swift Current Creek near Duncairn created a reservoir with a storage capacity at full supply of about 105 million cu. m. A dam across Rushlake Creek created secondary storage of about 12.3 million cu. m. The Swift Current main supply canal extended from a weir in Swift Current Creek for 33.6 km to Highfield reservoir, making water available to a large number of farms. The Herbert main canal extended from Highfield reservoir for 35.2 km, providing water for the Rush Lake and Herbert districts. (Spence, p. 91.)

But the biggest development of all was the one that had been dreamed of since the days of Palliser and Hind: a dam on the South Saskatchewan River.

New life for an old dream

New life was breathed into that dream in January 1944 when James G. Gardiner, the federal minister of agriculture, authorized test drillings to determine if the underlying soil and rock could support the massive weight of a dam. In a test drilling, samples were taken at carefully recorded depths. Moisture content, density, shear strength and permeability were among the parameters tested. If the depth of unstable material proved too great to support construction of a dam, the drilling was discontinued.

Meltwater from the massive glaciers that covered Saskatchewan during the last ice age carved the South Saskatchewan River channel. Today's relatively shallow channel only occasionally matches up with the original deep one, which over the millennia has filled with silt, sand and gravel, none of it fit to support a dam. Sometimes the old channel is on one side of the valley and sometimes on the other, but in test drillings it always turned up. Ultimately, the feasibility of any particular site depended on how far beneath the old channel lay solid bedrock, or at least firm shale—though there was some concern that the shale might not be firm enough.

During the Cretaceous Period, more than 70 million years ago, much of the Prairies was covered by the warm, shallow Bearpaw Sea. Large rivers flowed into the sea from the west, carrying clay and silt that settled onto the sea floor. Over time, that silt was compressed into a dark grey, flaky shale—itself perhaps not the most stable material on which to rest millions of tonnes of dam.

The site for the first test drillings, upstream from Saskatchewan Landing, was chosen because it was believed the river bottom might be more solid there, but the site proved unacceptable.

PFRA's engineers moved on to other sites suggested by consulting geologist (and professional engineer) Dr. John A. Allan of the University of Alberta, a recognized authority on the geology of the region. They investigated six sites in a 21-km stretch of the river, upstream from Pennant Ferry to 6.4 km upstream of Cabri Ferry. In the end, R.O. Peterson, the engineer in charge of foundation investigations, reported that at two of the six sites a dam of the required height could be constructed, but it "would be costly." (Spence, p. 125.)

Meanwhile, influential residents of Moose Jaw and Regina were

STOCK PHOTO

"A dam 85 feet high, and 600 to 800 yards long (a few miles lower down the length of the dam would be much less) across the deep narrow valley in which the South branch flows, below where the Qu'Appelle Valley joins it, would send its waters down the Qu'Appelle Valley, thence down the Assiniboine past Fort Garry, and thus establish a splendid and probably uninterrupted navigation, for steamers of large size, for a distance exceeding six hundred miles."

– Professor Henry Youle Hind

arguing against a dam in the Cabri area. They felt it was too far upriver to provide an economically feasible water supply for the two thirsty cities. They wanted to see a dam at the elbow—pretty much where Palliser and Hind had suggested one a century before.

Dr. Allan had misgivings; he knew that test borings by the Department of the Interior in 1913 and 1914 had shown unstable foundation conditions at the elbow and points upstream. Nevertheless, he next chose site 7, a short distance above the elbow, for a test drilling. Sure enough, drilling revealed a great depth of sand and other unstable material. Dr. Allan advised abandoning the site.

E.L. Gray, PFRA's superintendent of water development, pointed out that 9.6 km upstream from Outlook, outcrops of sandstone along the river banks might indicate bedrock near the surface, and Dr. Allan selected two more sites. At site 8 the drillers again found a channel filled with a considerable depth of sand, but also a number of sandstone "lenses" (deposits of sandstone thick in the middle and tapering off at the edges). Site 9, between the elbow and site 7, had a greater depth to bedrock than any site yet, and badly mixed layers of unstable material above it.

Right page (top): A geologist studies soil samples from the South Saskatchewan River Project site in early 1960 as construction gets under way. *(Agriculture and Agri-Food Canada # 60-02-012.)*

Right page (bottom): This map shows the location of the various sites where test drilling was carried out as engineers and geologists sought a suitable location for the South Saskatchewan River Project. *(From the design and construction of GARDINER DAM and associated works, p. 2. Courtesy Industry Canada. Reproduced with the permission of the Minister of Public Works and Government Services.)*

"I think we have found a site!"

And then, Spence recalled, "Early one morning, G.L. MacKenzie, Chief Engineer of the PFRA, was waiting for me at my office door. After a cheery 'Good Morning,' and without further ado, he exclaimed jubilantly, as he spread a Sectional Sheet on my desk, 'I think we have found a site!'" (Spence, p. 129.)

MacKenzie showed Spence the place downstream from the elbow where Coteau Creek joined the river. A low divide separated the current Coteau Creek channel from an apparent earlier course that had once joined the creek to the river 4 km further upstream. The resulting depression, MacKenzie pointed out, together with the current Coteau Creek Valley and the river on the east, formed the boundaries of a triangular tableland of about 300 ha, 4.5 m above the river.

"Then, laying his pencil down across the section of the river shown on the map, immediately east of the plateau, he continued, 'If foundation conditions are favorable for the construction of a dam in this section, the river can

STOCK PHOTO

The joys of irrigation

"Irrigation makes all the difference; with plenty of soft reservoir water it's a joy to see things grow—besides, to be able to pick delicious fruit, such as you see, off your own trees; to have your own crisp vegetables fresh out of the moist earth for your table; to have cut flowers in every room in the house, and bouquets for your friends and to give to people who are sick; that really does something for you! The PFRA have made it possible for us women folk to live on our prairie farmsteads, even in the driest years, without losing our souls." – A farm wife. (Spence, p. 81.)

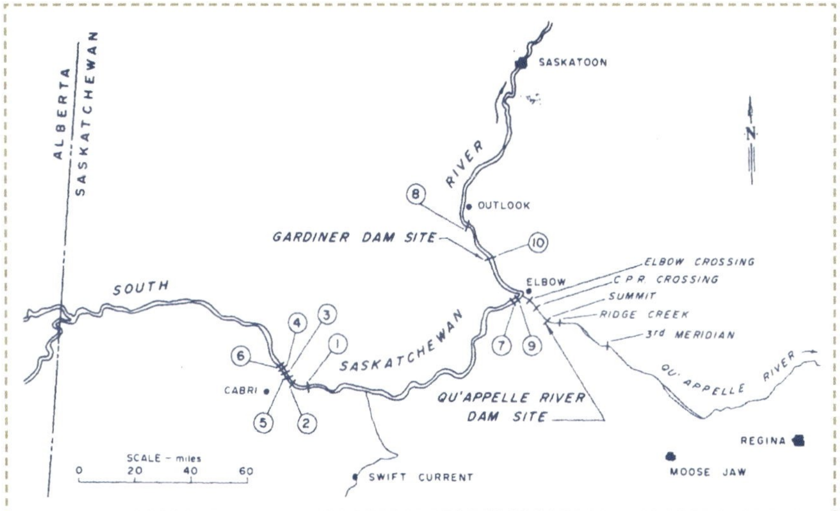

be diverted round the site by way of the Coteau Creek Valley, and the dam constructed in the dry.'" (Spence, p. 129.)

Dr. Allan examined the area and suggested another test drilling site, site 10. The previous most promising site, site 8, was also extensively drilled. Foundation conditions at both were similar, but the possibility of diverting the river through the Coteau Creek Valley gave site 10 the edge.

Spence consulted with Major General H.B. Ferguson of Washington, D.C., past president of the Mississippi River Authority and at one time in charge of flood control measures along the lower Mississippi River. Ferguson was working as a consulting engineer on another PFRA project, the St. Mary River Dam in Alberta. He concurred that the only feasible type of dam was an earthfill structure.

To learn more about the construction of earthfill dams in similar conditions, a party of engineers, including General Ferguson, MacKenzie, Dr. Allan, G.W. Parkinson, chief design engineer, R.O. Peterson, chief soil mechanics and materials engineer, and Reg Bing-Wo, senior design staff engineer, journeyed south to inspect the Fort Peck Dam on the Missouri River, then the world's largest earthfill dam: its reservoir had a storage capacity of more than 23.4 billion cu. m. They also visited the Garrison Dam, then under construction in North Dakota.

After a 1946 soil survey to judge the irrigation potential of the proposed dam, an interim report was prepared for the federal government. Dated May 30, 1947, it concluded, "Based on the results of these preliminary investigations, it can now be said with a fair degree of certainty that an earthfill dam can be constructed at either Site 8 (Outlook) or Site 10 (Coteau Creek)." (Spence, p. 137.) The projected cost: $66.4 million.

"A dam site surer and a dam site sooner"

After that flurry of activity, nothing happened until the appointment in 1951 of a Royal Commission to study the proposal. Its report was not entirely favourable, but the dam had strong champions in Saskatchewan Premier Tommy Douglas and in federal Conservative Party leader John Diefenbaker, who made the dam a campaign promise in the 1957 election. One of his Saskatchewan campaign brochures was titled "It will be a dam site surer and a dam site sooner if you re-elect John Diefenbaker."

In the election of June 10, 1957, Diefenbaker became prime minister of a minority government. On July 25, 1958, the provincial and federal governments signed a formal agreement providing for the construction of a dam on the South Saskatchewan River. Parliament approved the agreement on September 1, and an inaugural ceremony was held on May 27, 1959.

At the ceremony, attended by 15,000 people, Diefenbaker pulled a switch to mark the ceremonial beginning of construction. "Instantly, there was a loud blast, as half a ton of dynamite exploded and shot a mass of river-bottom earth mushrooming hundreds of feet into the still air," Spence recounted. "It was split-second timing, which gave the crowd, looking down from the hillside, a breathless thrill." (Spence, p. 151ff.)

Of course, real work had actually been under way for months. Beginning in the fall of 1958, roads were built to provide access to the site and work began on a construction bridge to allow equipment to cross the river.

In 1959, work started on the main dam. Dozens of large earthmovers began constructing the first two stages of the embankment, placing and compacting excavated earth on the east and west sides of the river, gradually narrowing the channel to half of its original width.

Work on the five river diversion tunnels, with a combined length of 6.4 km, began in 1960. The tunnels were dug by an electrically powered mechanical mining machine known as the mole, equipped with a rotary cutting head that could bore a hole 7.6 m in diameter. It scooped up its own cuttings and passed them to the rear to be hauled away in railway mine cars. Steel ring beams were installed around the circumference of the tunnel immediately behind the head of the mole at roughly 1 m intervals. Later the tunnels were lined permanently with concrete 76 cm thick. Three of the tunnels would eventually be used as hydro power penstocks.

Construction also began on high-level intake structures at the upstream end of the tunnels. Once the reservoir filled, water would be taken in through these structures at a point about 21.3 m above the bottom of the reservoir and fed into the diversion tunnels.

Work also proceeded on the control gates, raised or lowered to regulate the flow of water passing through the tunnels. The superstructures containing equipment for operating the gates appear as five cone-shaped houses on top of the dam—today among its most distinctive features.

The tunnels, intakes and control works were completed by the end of 1963. By then work was under way on the concrete spillway and on the Qu'Appelle River Dam 40 km southeast of the main dam. This dam, about 3100 m long and 27 m high, prevents the reservoir water from escaping into the Qu'Appelle Valley. A concrete conduit permits the controlled release of water.

In February 1964 the South Saskatchewan River was diverted from its original channel through the tunnels, allowing work to begin on the final stage of the main embankment. This stage alone required excavating, placing and compacting 26 million cu. m of earth and harvesting and placing almost 600 000 tonnes of rock riprap to protect the face of the dam.

While construction was under way, the site housed the largest implement workshop in Canada (measuring approximately 55 m by 40 m). That workshop serviced 49 earth buggies and 25 caterpillar tractors.

The creation of the large reservoir required major revisions to railway and highway routes between 1960 and 1965, including the rebuilding of one bridge crossing and the abandonment and removal of one combination highway and railway bridge.

Achieving closure

Building an earthfill dam typically involves bulldozing the fill across the river from both banks at the same time, slowly narrowing the river channel, until, in a last hectic hour or so, you achieve "closure."

George Spence was on hand for the closure of Gardiner Dam:

"Twelve giant bulldozers, six on each end of the fill, were engaged in bulldozing the loose material which had been previously piled into position on both banks of the river, into the rapidly narrowing gap in the cofferdam. It was imperative to fill the gap rapidly, as the swift flow was washing away much of the material being bulldozed into the river. In other words, the success of the operation depended entirely upon the rapidity with which the material could be bulldozed over the ends of the fill into the gap through which the water was flowing . . . It was, indeed, an unrelenting battle between the onslaught of puny man and the tremendous latent forces of nature. A compensating factor in the struggle was that while the velocity of the flow increased as the gap was narrowed by the bulldozing operations, the pond level above the fill was slowly but surely rising, with the result that more and more water poured into the diversion tunnels . . .

"The situation became more tense by the minute as the opening in the fill was being rapidly narrowed by the bulldozing operation, while the swift flow through the opening cut its way through in a narrow muddy boiling stream, carrying much of the bulldozed material with it down the river.

"Then, while the cameras clicked, four powerful bulldozers—two on each end of the fill—ground forward abreast pushing two great masses of loose material, which, in a passing minute or two, cascaded into the gap and landed in a great heap right into the narrow opening in the dyke. They sealed off, completely, every trickle of the flow. Closure had been achieved!

"I glanced at my watch and noted, for the record, that it was '1:53 p.m., February 14, 1964' . . . I turned to the Project Engineer, Walter B. Thomson, who stood beside me, and suggested to him that we should walk across the newly-created earth bridge. Consequently, we stepped out together, walking abreast to the other side of the channel.

"It can be said, therefore, that for the first time since the dawn of creation, two exultant humans, in the persons of Walter B. Thomson and this writer, walked across the mighty South Saskatchewan River Valley on dry land!" (Spence, p. 159.)

Building the construction bridge to allow men and equipment to cross the river, March 1960. *(Agriculture and Agri-Food Canada # 60-03-009.)*

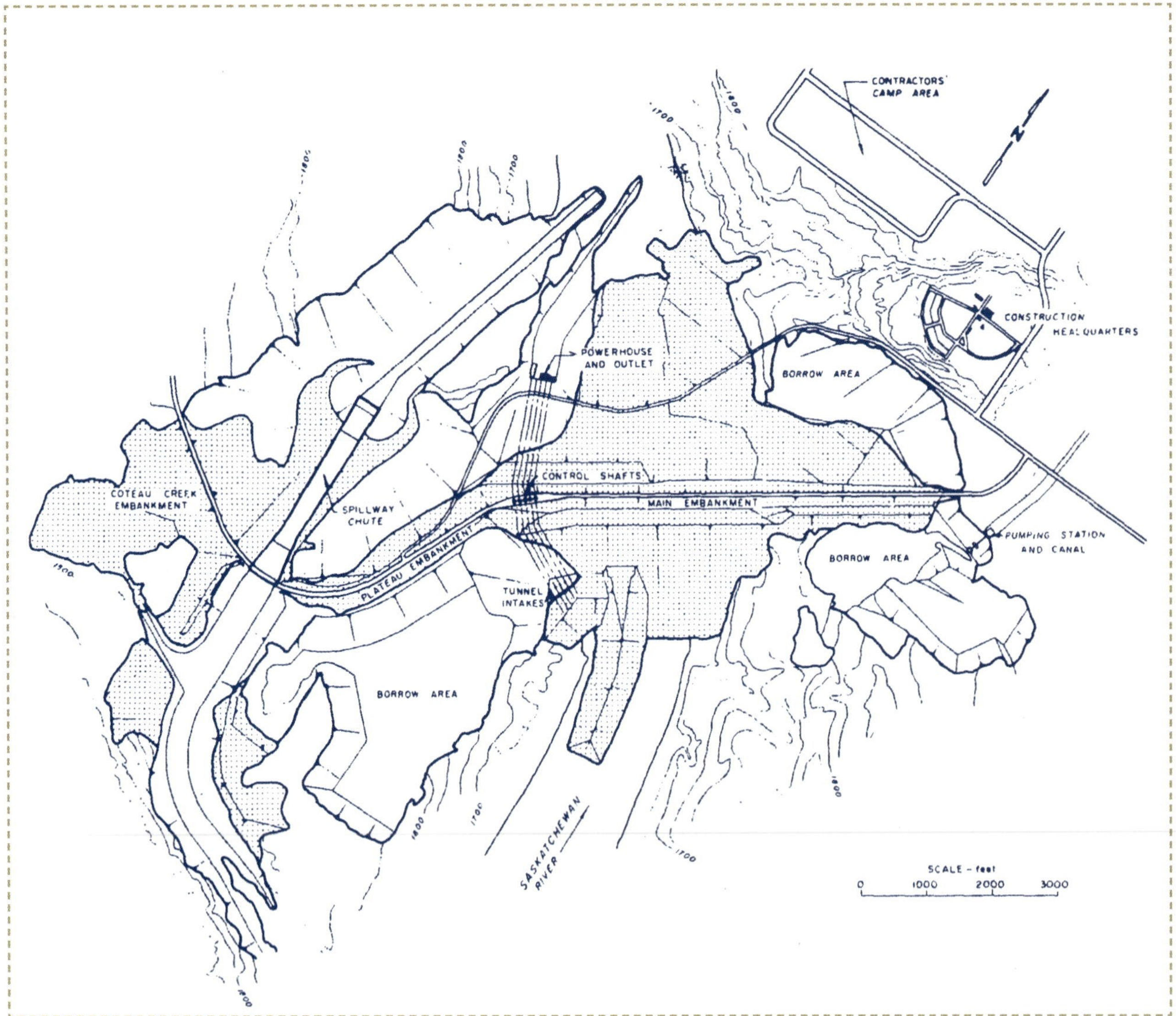

The final phase of construction of the massive spillway began in 1965. The spillway, almost 1.2 km long and capable of discharging 7500 cu. m of water per second, required 280 000 cu. m of reinforced concrete. When the impounding dams and spillway were completed in 1967, the reservoir was already well on its way to being filled for the first time.

Moving slopes

Engineering a large project like the Gardiner Dam always involves troubleshooting a series of problems, large and small. A major difficulty arose early on: as workers dug the deep excavations for the embankment foundations, some of the slopes moved. Despite all the

Left: This map shows Gardiner Dam's layout and principal features. *(From the design and construction of GARDINER DAM and associated works, p. 4. Courtesy Industry Canada. Reproduced with the permission of the Minister of Public Works and Government Services.)*

Right page: Giant steel ring beams await installation in the Gardiner Dam diversion tunnels, October 1961. *(Agriculture and Agri-Food Canada # 61-10-069.)*

drilling and analysis, the Bearpaw Shale in the area turned out to be weaker than expected.

The original design had to be revised to flatten the slopes of the main earth fills, thus spreading the load over an even greater area than originally planned. As a result, the base of the dam is nearly 2.5 km wide at the highest section.

The South Saskatchewan River Project (which includes Gardiner Dam and the smaller Qu'Appelle Dam, plus the resulting Lake Diefenbaker reservoir) was officially opened at a large public ceremony in 1967, as part of Canada's Centennial celebration. Ownership of the project was transferred to the province of Saskatchewan on April 1, 1969. PFRA continued to perform on-site duties until April 1, 1997, when the province took over the project entirely. It's now the responsibility of the Saskatchewan Watershed Authority.

The South Saskatchewan River Project

STOCK PHOTO

Qu'Appelle Dam Specifications
Length: 3100 m
Height: 27 m
Volume of earth fill: 10.4 million cu. m

Gardiner Dam Specifications
Earthfill Dam
Length: 5000 m
Height: 64 m
Volume of earth fill: 65 million cu. m
Spillway
Length: 1200 m
Width: 160 m
Discharge capacity: 7500 cu. m/sec

Lake Diefenbaker Specifications
Design Full Supply Level (FSL): 556.87 m
Effective drainage area: 126 000 sq. km
Length: 225 km at FSL
Maximum depth: 58 m at FSL
Area: 43 000 ha at FSL
Storage at FSL: 9.4 billion cu. m

Above (left): Work on the spillway, October 1964. *(Agriculture and Agri-Food Canada # 64-10-023.)*

Above (right): The Gardiner Dam control shaft superstructures, holding the equipment for lowering and raising the control gates in the diversion tunnels, under construction in May 1964. *(Agriculture and Agri-Food Canada # 64-05-016.)*

Right page: Workmen are dwarfed by the massive Gardiner Dam spillway. *(Saskatchewan Archives Board R-B4475-5.)*

The South Saskatchewan River Project today

Today hundreds of farmers use South Saskatchewan River water to irrigate a variety of crops. Water is pumped from the east end of the dam for the Outlook area and the Saskatoon South-East Water Supply Project, a series of five reservoirs connected by canals and pipelines that supply water to towns, three potash mines, and irrigation, waterfowl and recreation projects.

Further downstream, SaskWater extracts water from the river just south of Saskatoon for the town of Vanscoy, two potash mines, two golf clubs and several individual water supply systems.

The city of Saskatoon uses water from the South Saskatchewan River for domestic and industrial use, returning most of it to the river in the form of treated waste water.

Water from Lake Diefenbaker is also released through the Qu'Appelle River Dam to improve water supply and quality in the Qu'Appelle River system. (Otherwise Buffalo Pound Lake wouldn't have enough water to

Above (left): Former Saskatchewan Premier T.C. Douglas, Premier Ross Thatcher, former Prime Minister John G. Diefenbaker and Prime Minister Lester B. Pearson unveil a plaque at the official opening of the South Saskatchewan River Project on June 21, 1967. *(Agriculture and Agri-Food Canada # 67-07-051.)*

Above (right): A huge crowd gathers for the official opening of the South Saskatchewan River Project on June 21, 1967. *(Agriculture and Agri-Food Canada # 67-07-028.)*

Right page (left): Former premier and federal minister of agriculture James G. Gardiner (left), Premier T. C. Douglas (second from left), Prime Minister John G. Diefenbaker (third from left) and other dignitaries enjoy an outdoor lunch prior to the inaugural ceremony for the South Saskatchewan River Project on May 27, 1959. *(Agriculture and Agri-Food Canada # 59-05-031.)*

Right page (right): A 1989 aerial view of Gardiner Dam. *(Agriculture and Agri-Food Canada # 89-07-341.)*

meet the needs of Regina and Moose Jaw and two major industrial plants at Belle Plaine, the Mosaic potash mine and the SaskFerco fertilizer plant.)

All of these "consumptive uses" take only about 5 per cent of the South Saskatchewan River flow in an average year. At least that much again is lost in evaporation from Lake Diefenbaker's surface.

The remaining water, 90 per cent of the river flow, is released through the turbines of the Coteau Creek Hydroelectric Station at Gardiner Dam to produce electricity and maintain the downstream flow at a level adequate for the needs of ferry operators, irrigation farmers, people enjoying recreation on the river and various municipal and industrial users.

Juggling the needs of the many different users of the South Saskatchewan River, upstream and down, is the biggest day-to-day

operational challenge of the South Saskatchewan River Project. The challenge is greater in dry years: the South Saskatchewan River Project can mitigate the effects of drought, but it can't conjure water out of nothing. Ultimately, nature is still in control.

"Worth every cent"

Still, there seems little doubt that the South Saskatchewan River Project, the biggest single engineering project ever undertaken in Saskatchewan (its final cost of $121 million would be equivalent to around $1 billion today), has already proven its worth, even though plans are constantly afoot to utilize it even better.

A Canadian government study in 1988 noted that "The project has brought land, water and human resources together to provide a productive and aesthetically pleasing environment that would otherwise not exist. This enhances and increases conservation projects of plants, wildlife, waterfowl and fish habitat." (Macdonald, p. 99.)

Brad Fairley wrote in the 1990 federal government review of the South Saskatchewan River Project that "there have been some very bad dams built (but) construction of the Gardiner Dam was one of the smartest things the province ever did." (Macdonald, p. 99.)

Perhaps the final word, though, should go to the anonymous woman quoted by Max Macdonald: "Living in a community which hitherto lacked the benefits of indoor plumbing, she noted pithily, 'Any project that keeps my tender bottom off the seat of an outdoor privy at forty below is worth every cent no matter what the cost.'" (Macdonald, p. 99)

CHAPTER 2 SOURCES

Books

Engineering Service, Prairie Farm Rehabilitation Administration. *the design and construction of GARDINER DAM and associated works.* Ottawa, 1980. Government of Canada.

Macdonald, Max. *The Dam the Drought Built: A History of the South Saskatchewan River Project.* Regina, 1999. Canadian Plains Research Center.

Spence, George. *Survival of a Vision*. Ottawa, 1967. Canada Department of Agriculture.

Other publications

The South Saskatchewan River Project. 1967. PFRA.

The South Saskatchewan River Project. 1986. Government of Canada, Agriculture Canada PFRA and Saskatchewan Water Corporation.

Websites

Agriculture and Agri-Food Canada, PFRA Online. *Lake Diefenbaker and How It Operates.* (March 15, 2006.) <http://www.agr.gc.ca/pfra/pub/lakedb_e.htm>

Saskatchewan Watershed Authority. *South Saskatchewan River Project Fact Sheet.* <http://www.swa.ca/publications/documents/FS-308.pdf>

Saskatchewan Watershed Authority. *Water Control-Saskatchewan Dams and Reservoirs.* (March 15, 2006.) <http://www.swa.ca/WaterManagement/DamsAndReservoirs.asp>

STOCK PHOTO

CHAPTER 3

You *Can* Get There From Here

"You've got to move things around."

Saskatchewan became a province, the majority of its traffic was horse-drawn: in 1906, Saskatchewan boasted only 22 registered motor vehicles.

By 1911, Saskatchewan had 821 registered vehicles. A Model T Ford sold for $1,025 (roughly equivalent to $20,000 today). The next year the provincial government appointed a Board of Highway Commissioners, whose 1912-1913 Annual Report took eight pages of tiny print to list all of the "highway improvements undertaken by Government Road Gangs" that season and 25 bridges either built or under construction. (Board of Highway Commissioners, 1913.)

All that road building was just in time: in 1916, new methods of mass production dropped the price of a Model T to $545, around $10,000 in today's money. Demand

Above: Early work on the Candle Lake Road, 1946. *(Saskatchewan Archives Board R-A5515.)*

Right page: In an agricultural province like Saskatchewan, good roads were needed early, to avoid situations like this at Norquay. *(Saskatchewan Archives Board R-A2260-2.)*

Building the Roads

"You can't get there from here," goes the old joke, but in Saskatchewan, you usually can, thanks to the most extensive road system in Canada, totalling more than 190 000 km.

Building, improving and maintaining those roads is a major undertaking. While the peak years of road building were the 1950s, roads were of concern from the very beginning of the province.

From cart trails to the first highways

Saskatchewan's earliest roads were cart trails that connected trading posts, forts and missions in the 19th century. Even after

Early paving

The route from Regina to Pense, later part of the Trans-Canada Highway, was one of the first paved roads in Saskatchewan. The 1941 Annual Report of the Department of Highways described in detail the first stage of the process:

"The width of the right of way was increased from 66 feet to 150 feet in order to secure by side borrowing the material required for the construction of a subgrade having an average height of two feet above natural surface elevation with a crown width of 32 feet and with side slopes of four horizontal to one vertical. The material in the embankment was placed in layers not exceeding eight inches in depth, and each layer was consolidated at optimum moisture content by means of sheepfoot rollers. On the completed subgrade a six-inch subbase of plant-mix waterproof stabilized gravel was constructed. The surface of the subbase was seal-coated as a protection against the abrasive action of traffic, pending the laying of a bituminous wearing surface during the next construction season . . . "

The report goes on to note that "a high standard of construction was required on this route to meet the demands of present day traffic, but it is not practicable for the present population of this province to finance the construction of a road of like standard throughout the limits of the present provincial highway system." (Department of Highways, 1941.)

for automobiles surged. In 1917 the Department of Highways replaced the Board of Highway Commissioners. Highway development, however, was modest through the 1920s. In 1927 there were still only 4200 km of provincial highway (though many more kilometres of rural roads, of course).

Highway work essentially stopped during the Great Depression. Highway traffic also dwindled, as the number of motor vehicles declined from 128,000 in 1929 to 91,000 in 1934.

Grading more important than paving

The first hard-surfaced roads appeared in the 1930s, but in 1931 fewer than 3 per cent of farms even had gravel road access, so any roadwork focused more on creating adequately graded and gravelled roads than paving existing ones. By 1944 Saskatchewan still had only 222 km of hard-surfaced roads. (*Encyclopedia of Saskatchewan.*)

The key influence on Saskatchewan's rural road structure, at least in the South, was the grid system of land surveys. The survey divided the province into 36-square-mile townships (6 miles on a side), which were further divided into 36 sections, each 640 acres (1 square

Above: The first cut for Highway 33, August 1954. *(Saskatchewan Archives Board R-A13310.)*

Right page: Laying the blacktop on the Trans-Canada Highway west of Swift Current. *(Saskatchewan Archives Board R-B6300-1.)*

mile). Over most of the area, there was a 66-foot road allowance every mile on the east-west line and every 2 miles on the north-south line. Southern Saskatchewan is thus criss-crossed with "grid roads" that follow these road allowances.

Tom Gentles joined the Department of Highways in 1947 as a resident engineer. He went on to be a supervising engineer and design engineer in various locations, and spent his last nine years with the department, from 1972 to 1981, as deputy minister. He said construction of the grid road system was gradual. "Where there was a railway going through and they needed transportation to get to the elevator, they developed these road allowances. When I started, most of them were just trails . . . The farmers were just using them to access various parts of their quarter sections or half sections,

STOCK PHOTO

Finding gravel

Gravel comes from a gravel pit. Saskatchewan has plenty of gravel, thanks to the ice age glaciers that ground across it, but finding gravel when and where it's needed isn't always easy.

"Often the engineers out there had to explore . . . ask around, and find a likely looking site," Gentles said. "I remember once testing a gravel pit all in one day.

"It was near Churchbridge. We needed gravel, didn't have any. Word got around that there was a certain place there, so I ended up contacting everybody in Churchbridge who had an able-bodied man. We bought about 20 shovels—cleaned out the hardware store . . . They could dig one of these 6-foot test pits in one day. So they dug it out and proved that the gravel was there.

"I got back two shovels at the end of the day. It was part of their pay."

and it was just the main roads the farmers were using for hauling with horses and wagons that were sort of halfway graded up.

"If you go back into the '30s the grading was done with Fresno scrapers and a team of horses. I remember . . . my dad in 1932, working off his municipal taxes. I can show you the old trail yet that he was working on that day, with all the other farmers who couldn't afford to pay their taxes, with a team of horses and a Fresno scraper."

Road construction boomed in Saskatchewan in the 1950s as society changed, traffic increased, and more and more goods were shipped by heavier and heavier trucks. "Transportation is an essential component of the whole economy," Gentles pointed out. "Particularly with the farming situation that we have here, spread out as it has been, you needed this network of roads, running from just prairie trails . . . (to) the major highway system."

The earliest main roads, like the grid roads, were built on the square. "They just felt that was the way to

Above: Road testing in 1956. *(Photo courtesy Saskatchewan Highways and Transportation.)*

Right page: Jets perform a flypast over the crowd at the official opening of the Saskatchewan leg of the Trans-Canada Highway, August 21, 1957. *(Saskatchewan Archives Board R-A11435-8.)*

go," said Gentles. "If you wanted to go 6 miles up from the railway track you didn't think of following the railway track, you went on the road allowances." But the roads of the 1950s that Gentles and other engineers designed were more likely to follow the railroads. "We just started to . . . realize the economics of highway travel, if we can shorten the distances," Gentles said.

Putting it together

In the early years, people on the ground chose road routes. In the late 1940s and especially during the 1950s and into the 1960s, aerial photography began to play a key role in locating new roads (more about that later in this chapter).

The route was surveyed and the centre line determined and marked. After that, cross-sections of the topography were created at regular intervals. A ground profile was plotted, then a road template placed on top of the profile. From that the amount of cutting and filling required to make the road fit its design requirements could be determined. In the 1950s, a typical road would have a 24-foot-wide roadway, 3:1 side slopes, a 15-foot ditch, and then a 3:1 back slope.

Next, survey crews staked the ground, the stakes telling the contractor exactly how much dirt he'd have to excavate or fill in. It was labour-intensive, Gentles said. "I have to confess, sometimes . . . (when) one engineering crew's got three projects going, and all the contractors are crying for stakes . . . you just, from the cross-sections, figure out the cut stakes and the fill. And then you just go out with your notes, no instrument . . . and put these stakes in. You can go down and do 5 miles a day without any problem. That will keep the contractor going for five weeks probably, and give you a chance to survey some of the other ones."

Such shortcuts were sometimes necessary because of the amount of work being done in those boom years. "Having two projects was very commonplace. Every once in a while you'd get a break and there's only one road to do. That's a holiday."

Once the cutting and filling was done, the road would be packed using a "sheepsfoot" roller (a big metal cylinder with protruding rectangular teeth). The road would then have to be checked again (especially if it was going to be paved) "to make sure the contractor has it right," Gentles said. "If he hasn't got it right then you put in blue tops. Say he needs an extra 2 inches somewhere, you pound this thing in down to the level that the contract required. Then the contractor comes in—they just love these, of course—

(and) puts 2 inches on what he hoped was a finished road."

On a gravel road, the next step was spreading the gravel. "We used to figure 1,000 cubic yards per mile," Gentles said. "We didn't want to put it on all at once, because it made it pile up too much, so we'd put in 500 or 600 (cubic yards), let it settle in, then stockpile the rest of it. Let maintenance bring that on just when it was needed."

Paved roads had to be wider than gravel roads. The exact width would depend on how thick the base course needed to be, and that depended on "the strength of the earth, clays being the weakest, (and) the Regina gumbo pretty well the weakest there was," Gentles said.

"If it's fairly strong and it's not Regina gumbo you might get away with 6 inches of base course. If you want a 24-foot-wide road and you want 6 inches of base course, you have to build the subgrade wider than that. The width of the subgrade is designed to handle the paving thickness that has to go on top of it and still maintain the slope. . . . That could vary quite a bit."

The main element of the base course was a good, graded gravel, not uniform, with the biggest pieces being no larger than about five-eighths of an inch in diameter. It also included a bit of silt, "because what you want is something very, very dense, no spaces or holes or anything else," Gentles said.

The thickness of asphalt that went on top depended on the amount of traffic. "It could be just an inch and a half, one lift, one shot from the asphalt mixer. If it's the Trans-Canada or some of the other big highways they'd probably put on two lifts, 3 inches."

A dust-free road by any other name . . .

And then there were the "thin-membrane" roads, originally referred to as "oil-treatment" roads: three-quarters of an inch of bitumen oil laid on top of a good base of clay. "Oil treatment wasn't a good word for it," Gentles said,

First with the Trans-Canada

One of the most significant accomplishments during the 1950s was construction of Saskatchewan's segment of the Trans-Canada Highway. The province signed an agreement in 1950 to provide funding for the road, which would parallel the Canadian Pacific Railway main line. Just seven years later, on August 21, 1957, Premier T.C. Douglas officially opened Saskatchewan's 653-km east-west stretch of the highway, which reduced border-to-border driving distance in the province by 89 km and eliminated 23 level railroad crossings. Saskatchewan was the first province to complete its section of the national road, constructed to the same standards nationwide.

STOCK PHOTO

but added, "Thin membrane isn't a very good word for it either.

"We got a lot of static from those because you build the darn thing and a few months or a year later they need a lot of patching," he said. "But they could be done for about one-tenth of the cost of a fully designed paved road.

"We'd get accused of not knowing how to build a road, but we'd say, 'Okay, the engineers have no trouble building a road that won't fail, no problem at all, but the only thing is with a limited amount of money we can only build one-tenth as much. So do you want to wait 10 years, or do you want this oil treatment?'

"I don't think we ever really sold people on it, but we were convinced because we would look at the economics of travel, and the cost of travel on these oil treatments was so much less than the gravel roads. Maintenance was a little bit higher, but even counting the extra maintenance, we saved.

"I'm still convinced it was the right thing to do. I don't think a lot of roads in Saskatchewan would be paved now if it wasn't for those."

Without the development of oil-treatment/thin-membrane roads, Saskatchewan would probably never have been able to claim the title of the province with the most paved roads in Canada. According to Allan Widger, Executive Director of the Engineering Standards Branch for the Department of Highways and Transportation, oil-treatment/thin-membrane roads served their purpose of "trying to get everybody out of the mud." And, he added, "they still perform well under reasonable traffic loads. We still have pretty nearly 7000 kilometres of them in the system." However, no more are being built.

By 1960, Saskatchewan's total highway mileage had swelled to 8,370.77 miles (13 393.23 km), of which 154.16 miles (246.66 km) was earth subgrade, 5,907.86 miles (9452.58 km) gravel surface, 1,585.02 miles (2536.03 km) bituminous pavement, 591.75 miles (946.8 km) oil-treated ("thin-membrane"), and 98.87 miles (158.19 km) not improved. (Department of Highways, 1960.)

Today the highway network consists of 26 250 km of roads, roughly 14 per cent of the entire rural road network of more than 190 000 km. Fully 51 per cent, or 13 460 km, is paved, 26.5 per cent is thin-membrane, 22 per cent is gravel, and 0.5 per cent is ice road.

Roads are classified by social, economic and connective function. About 38 per cent of Saskatchewan highways are classes 1 or 2, which have the highest traffic volumes: they connect major cities and regional service centres that have populations greater than 1,000. Classes 3, 4 and 5, into which the remainder of the highways fall, link

Four lanes are better than two

The next big improvement (beyond simple paving) was four-lane highways. The first four-lanes in Saskatchewan, completed in 1961, ran from Regina to Lumsden and Regina to Balgonie.

Before building four-lane highways, engineers had to decide how wide the median should be. "We knew that if you were driving at night on a two-lane highway or something like that and somebody is coming toward you with bright lights on, it's a little disconcerting," Gentles said. "As design engineer I went out into the fields, into an area that's now just a couple of hundred metres south of the Golden Mile Centre (a Regina shopping mall) . . . with a counterpart from the Department of Highways. We'd come at each other with bright lights on and then gradually increase the distance until we decided what width we needed between the lanes of a four-lane highway . . . The width of the four-lane is still built on that 1954 research."

Left page: Building a new road still begins with surveying the route, just as it did half a century ago.

Above: An impressive lineup of some of the heavy equipment used today in Saskatchewan roadwork. *(Photo courtesy Saskatchewan Highways and Transportation.)*

Above: A Caterpillar Roto-Mill rips up and recycles old asphalt. *(Photo courtesy Saskatchewan Highways and Transportation.)*

Right page: A Brandt Power Unit pulls a car loaded with heavy equipment along a rail line. *(Photo courtesy Brandt Industries Ltd.)*

communities with populations of fewer than 1,000 or give access to large parks and industrial sites. Classes 6 and 7—non-highway roads—have the lightest traffic, and provide access only to individual residences, farmland, small parks and industrial sites or other properties. (*Encyclopedia of Saskatchewan.*)

The challenges—and research—continue

Saskatchewan's road builders and maintainers face the same basic problem today as they always have. As Widger put it, "You tend to need to have to build a lot of roads to serve a small, scattered population, because people expect the same level of service whether there's 500 vehicles on a road or 2,000 . . . Probably our biggest challenge is we've got a million people and just not enough revenue in the province to spend on such a vast road system."

To try to meet expectations in as economical a way as possible, the department conducts a lot of research and continually looks for new and better ways to do things. For example, Widger said, "Currently we're working with satellite imagery, space shuttle radar imagery, all of that kind of thing to replace the air photos we used for the last 50 years."

Computers have opened up new possibilities. "We're getting into real-time monitoring of things like our slope instrumentation and our frost thermistor sites, where we measure the rate at which the frost goes in and out of the ground," Widger said. The department has been monitoring the movements of slopes for 40 years, but could only check slope indicators three times per year at most. "Now finally we have the capability of monitoring them in real time so we can actually start answering some of the questions we've been struggling with for the last 40 years as to how things are actually moving, whether it moves all at once or whether it's a sequential thing up the slope."

Computers have also made possible a remarkable asset management system. Specially equipped vehicles, travelling at highway speed, use laser technology and infrared cameras to scan and photograph roads. By measuring rut depths, the amount of cracking, and more, they build up a digital profile of the kind of ride users of the road experience. "We collect about 3 terabytes of data a year," Widger said. "This is our third year of running it . . . the first year we collected more data in three months

The Road/Rail Unit

In 1984 the Railway Transport Committee of the Canadian Transport Commission conducted a branch line inquiry. During the inquiry, Saskatchewan suggested that an alternative to branch line abandonment or traditional short-line railway operations would be to develop a road/rail operation.

Five years later, in the autumn of 1989, Southern Rails Cooperative Ltd. (SRC) began just such an operation, under operating authority from the province, on Avonlea subdivision, a branch line previously owned and operated by Canadian National that served Parry and Truax country elevators.

The Saskatchewan Road/Rail Operation Project Evaluation Final Report described the operation this way: "The SRC road/rail short line utilizes a single modified high-rail highway tractor capable of serving a number of separate branch lines, which need not be contiguous. The operation involves loading of rail cars at country elevators and delivery of cars to the main line carrier at an interchange point. The road/rail power unit travels between elevator points as a railway unit over retained railway lines, and moves between branch lines as a highway unit. The road/rail power unit presents an application of new technology utilized in an innovative, practical manner." (Travacon Research Ltd.)

The road/rail power unit looks like a truck, but has both tires and rail equipment. It can travel on the road at normal highway speeds, then convert to rail service in just three minutes at any 9-m crossing, after which it can pull or push 15 loaded 90-tonne grain cars at more than 60 km per hour (kph).

The unit was developed by employees of the Department of Highways and Transportation and built by Brandt Industries of Regina. Beginning in the early 1990s, Brandt Industries had a patent licence agreement which allowed it to exclusively manufacture and sell the road/rail power units, with the province receiving royalties from each unit sold. In just nine years, Brandt manufactured and sold 46 units (from which the province received $1 million), and also developed other product lines related to the road/rail business. In 2002, the Government of Saskatchewan sold its road/rail patent to Brandt Industries for $650,000.

Today versions of the Brandt Power Unit continue to be built in Regina and marketed around the world—although, ironically, not for the road/rail grain operations for which the power unit was invented. Instead, it's used primarily for "maintenance of way" operations (railway tie distribution and pickup, moving workers, etc.) or "emergency response" operations (i.e., moving rail cars in an emergency situation or getting equipment to the scene of a derailment).

with this piece of equipment than the department had created in its history before that."

The rubber does more than just meet the road . . .

New paving techniques are also being tested. In 2005 the department built a rubber asphalt demonstration project on Highway 11 just south of Chamberlain. By digesting rubber crumb into the asphalt, Widger said, "you get all the properties of the rubber in the asphalt: rather than it getting stiff in the winter and super soft in the summer . . . it's more of a constant performance through the year. (It) also adds a lot of durability."

Will we be seeing more rubber asphalt? Well, says Widger, "Initially, the economics look good. The biggest problem we have in Saskatchewan is our sources of aggregate. You need a good-quality aggregate. The aggregate is more expensive, but you don't put on as thick a lift, so it actually balances out in the long run . . . And then if we get longer performance, obviously that's where the economics comes in."

Right page: Today, remote sensing is based on satellite photographs, like this one of Gardiner Dam, as well as on aerial photographs. *(Photo courtesy Jack Mollard.)*

A purely Saskatchewan innovation is strengthening subgrades with cement or cementaceous products, "again to try to get away from the need for aggregates in areas where there are long hauls," Widger said. "We've had them out for five years and performance is looking good."

Road design has had to evolve with the times, as well. "The size and configuration of trucks . . . (have) changed significantly, and now we have to design the highways both from a structural and a geometric point of view to carry them," Widger said. "There's increased loading, so obviously your pavement structures end up being thicker. You have to deal with the turning movement and the tearing movements of the tridem axle groups, so that puts more strains on the asphalt surfaces or your sealed surfaces . . . Their turning radiuses and things like that are different than the old truck configurations, so you have to change your geometrics for things like interchanges and intersections to allow for it. Even things like the median widths in four-lane highway systems: we're widening them out now at major intersections so that there's enough room for a combination vehicle to park in the median as it's crossing, so it's not hanging over the lane."

Road equipment is now larger and more sophisticated. Widger mentioned recycling equipment that can grind up 150 mm of hot mix and the stuff underneath it, mix it all and lay it down again in a single pass. "It opens all-new opportunities for looking at things like foamed asphalt, where you can tear up the whole existing structure . . . you add hot asphalt with a little bit of water and you get an explosion, you create a huge amount of . . . asphalt steam, and so you can coat a lot of material. That way you can get the material so it's lubricated and compacted down without adding a large amount. You can lay thick lifts and you don't have a lot of liquid in there that has to cure out or, in the case of emulsions, evaporate out. That technology is moving ahead quite rapidly."

Roads are important

It's all a far cry from the early days of private operators improving their own roads with a horse-drawn grader. But in Saskatchewan, as both Widger and Gentles emphasized, roads matter.

"We need the road system because Saskatchewan is such a big producer of raw materials, which tend to be bulk commodities, and they all go to export," Widger said. "You're looking at taking a low-cost bulk commodity and getting it moved to market in the most economical way possible, so you're looking for good roads, and as heavy a vehicle as you can get."

Or, as Gentles put it, "Transportation is an essential component of the whole economy. You've got to move things around."

The View from on High

Jack Mollard and Remote Sensing

The first step in building a road is choosing a route. The best way to choose that route is from aerial photographs—and no one has more experience at that than Jack Mollard, president of J.D. Mollard and Associates in Regina.

Before the advent of aerial photography interpretation, all road routes had to be picked out on the ground. That isn't tough in wide-open prairie, but in the North, it's a different story.

"On the ground, a road locator takes off and wonders, 'Where the hell should I go?'" Mollard said. "It's not like the northern trapper, who makes a beeline straight through the bush to his trapline. You have to be able to view the terrain miles ahead, recognizing and outlining good (desirable) and bad (undesirable) ground, then try

"My thing is to look at a piece of the Earth from above and say, 'What is it, and how will it behave for a given use?'"

Above: Before aerial photography was widely used to help locate road routes, the work had to be done on the ground by road location parties like this one, reconnoitering the route between Montreal Lake and Lac la Ronge in 1939. *(Saskatchewan Archives Board R-A5860-1.)*

Right page: Jack Mollard (in toque) and associates in northern Manitoba in 1957, looking for a source of gravel originally identified from the air. *(Photo courtesy Jack Mollard.)*

to figure out how to avoid the bad ground and follow the good ground on a fairly direct route."

Tom Gentles worked as design engineer on several northern roads. He recalled hearing a story from fellow engineer Max Jacoby about locating the highway on the east side of Montreal Lake during the Second World War. "They had all these conscientious objectors, and they were the ones that got stuck with locating this road," Gentles said. "It was mostly marsh for mile after mile after mile. He said they would just have a sharp stick or stake and they would walk in a straight line, and as long as they could feel bottom with this stick they would keep on that route, but then when they could no longer feel the bottom, they would back off a little bit and start probing a different direction until they could find another tangent where there was bottom."

Working like that, Mollard said, you can find a route, but you can never be certain you've found the best route. "On the ground, you don't know what you've missed," he said. "You don't know where the good and bad areas are located, and you can easily miss the better ground."

When he looks at a set of stereoscopic aerial photographs, Mollard said, "I'll circle an area and say, 'that's rough rock,' circle a muskeg area and say 'this muskeg has ground ice (permafrost) in it . . . and that's bad news,' and then try to avoid both in selecting a road route."

Changes in the process

Remote sensing tools are improving, says Allan Widger of the Department of Highways. "Currently, we're working with satellite imagery, space shuttle radar imagery . . . (which) is good at establishing elevations, so you can actually drape the landscape over top of an air photo, and if

you're looking at things like gravel, or if you're looking for highway route locations, it can give you a rough profile. The big advantage of it is that you don't have to worry about tree cover, because it shoots through the trees."

Of course, choosing highway routes is only one use of remote sensing (of which aerial photography is just a part). Another highway-related task, as Widger mentioned, is finding gravel or aggregate. But remote sensing can also be used to find water, mineral resources, dam sites, old coal mines, hazards like potential landslides and contaminated land, and on and on and on.

Mollard said the whole process of road route selection has changed in the half-century since he began finding road routes in the North. For the last 35 years he and his brother, George, have worked together locating northern roads. "We've located nearly all the existing highway routes in northern Saskatchewan and Manitoba, the Northwest Territories and Nunavut," Mollard said. "In the early years we tried to find the most direct and economical route to build. Today our job is more challenging. Right now we're selecting a road route from Rankin Inlet, Nunavut, south to three alternative destinations in northern Manitoba—all three roughly 1000 kilometres long, each with a large variety of environmental controls: traditional habitats of polar bears and beluga whales, caribou calving grounds, provincial parks and heritage sites as well as other kinds of protected land. Even though there's a preferred route from engineering design, construction and cost, we may not be allowed to follow that route for environmental, social, cultural or political reasons. In today's world these concerns require careful assessment and decision making."

The projects—even just the kinds of projects—J.D. Mollard and Associates have worked on are too numerous to list here (they number more than 5,000). Mollard once calculated he had used the same set of aerial photographs for a relatively small area near Estevan for more than 20 different kinds of projects, from locating the Boundary and Rafferty dam sites to finding gravel, exploring for oil and mapping cave-ins over abandoned coal mines (a hazard that had previously taken a child's life).

For Mollard, it all boils down to "interpreting the landscape and its origin: how it was formed, what it's made of, and how it will behave, whether it is suitable or unsuitable" for a particular use.

Distinguishing features

Mollard has taught applied air photo terrain analysis to engineers and geoscientists throughout his career. When students ask him, "How can you possibly identify soil and rock formations from air photos?" he often replies that in the beginning he tried describing their distinguishing features: topography, drainage, erosion, land use and vegetation. "Then I say, to hell with

Above: This small-scale (1:130,000) stereotriplet illustrating many different landforms and modes of sediment deposition within a small area in Waterton National Park, Alberta, is one of the examples Mollard uses when teaching remote sensing. *(Photo courtesy Jack Mollard.)*

that approach, it's too much of a stretch. A better approach is identifying earth features in much the same way that you recognize friends and acquaintances. When you see sand dunes and volcanoes several times in 3-D air photos, you begin to recognize their characterizing visual appearances. And you can call them by name, just as you do friends.

"How do you recognize people? They've all got eyes, nose and mouth. You see them, you say, this is (someone I know)," Mollard said. "There are typically fewer landforms in the world than one's friends and acquaintances—maybe 100 or so basic landforms with recognizable minor variations. For me, the landforms I see almost daily in stereoscopic air photos are as easy to recognize as friends.

"But what is necessary to infer their properties is having prior knowledge of the geological history, structure and composition of landforms from correlating their appearance in air photos with that on the ground, whether the task is finding groundwater for a town, gravel for a rural municipality or a landing strip in the Arctic."

Seeing things in black and white

Mollard prefers to work from black-and-white panchromatic aerial photographs. "There are 256 grades from white to black, and their variations in pattern and intensity of light and dark tones help to characterize the many different kinds of soil and rock landscapes that aid their recognition," Mollard says. "I look for changes in grey tone that are meaningful, but often very subtle. Gravel deposits, for example, are usually lighter tones than clay deposits. As a further example, a glacier-deposited hilltop that's been eroded by wind and water for 10,000 years usually has very little organic topsoil left and its surface looks whitish after being plowed, while a nearby slough bottom looks darkish because it contains dark organic topsoil that's been washed down from the eroded hilltop over 10,000 years, along with the added growth and decayed slough-bottom vegetation over the same time.

"For me, true colour and false colour (infrared) air photos tend to mask subtle yet informative recognition detail associated with different soil and rock landscapes," Mollard said. He admitted, though, that they can be useful for identifying things like sediment in lakes, different types of trees and different kinds of land use.

New technology is making possible new ways of using aerial and satellite photography, Mollard added. He recently consulted on a 300-km pipeline across the Peruvian Andes. Troy Zimmer, a computer/ground-imaging satellite specialist in Mollard's office, obtained the precise pipeline route and a computer-generated digital elevation model covering the narrow pipeline corridor across the Andes. Zimmer then "draped" a high-resolution colour satellite image over the elevation model.

"I expected to see a two-dimensional image of a pipeline right-of-way cutline on the computer monitor when Troy called me and said, 'Would you like to fly along the route?'" Mollard said. "I said, 'Sure,' and was amazed at what I saw, sitting there in an armchair looking at the monitor and flying over the route: mountain slopes, pattern of erosion, landslides, even narrow trails."

Mollard recalled how decades earlier, flying in a helicopter over pipeline routes he had located in

Ask yourselves these questions:

When interpreting 3-D aerial photographs in his short courses, Mollard asks students questions like these:

What's the name of that feature?

What are the identifying characteristics that tell you how it was likely formed? Has it been shaped by the wind, water, ice or gravity? How can you tell?

What five meaningful things can you say about the feature? Is it a landslide, sand dune, volcano, or something else?

You're looking at a ground surface that's 10,000 years old. What do you think the scene looked like back then? How has it changed?

Imagine you're looking down at Regina and surrounding area. Imagine a half-kilometre of glacier ice over the city. What do you think happened when it melted? Would you agree that it formed a huge lake? Imagine 60 feet of glacial meltwater at the SaskPower Building. Also, imagine the height of waves striking a sandy lake shore as they approached Pilot Butte and White City.

Now drill two holes to 20 feet. What kind of soil are you likely to encounter at the TV tower east of Regina, where the water was deep? At Pilot Butte and White City?

How will these different landscapes behave for a given use? Will they behave differently if used for growing wheat? As the basement of a house? As a nine-hole golf course?

Canada's High Arctic, he would sit beside the pilot, pointing out varying terrain features to construction cost-estimators sitting in the back. "On the intercom I'd be instructing the pilot to 'Get over! Get down! Now higher, I can't see the route right-of-way!' And here I am, sitting in Troy's office, and unconsciously shifting up and down in my chair just as I did in the helicopter, directing Troy to 'Get down! You're too high! Lower!' Then I said, "Stop! Enlarge the scene! More! Even more!'" Mollard said this technology allows him to interpret landscapes virtually anywhere in the world.

"So much of our work is finding things," Mollard said. "An R.M. (rural municipality) says, 'We've run out of gravel. Where will we look for gravel?'" Over the years, he said, his brother George has searched for and mapped gravel for 80 R.M.s in the province.

"We've also searched for groundwater to install water and sewer systems for nearly 200 towns, villages and industries. My job involved examining the air

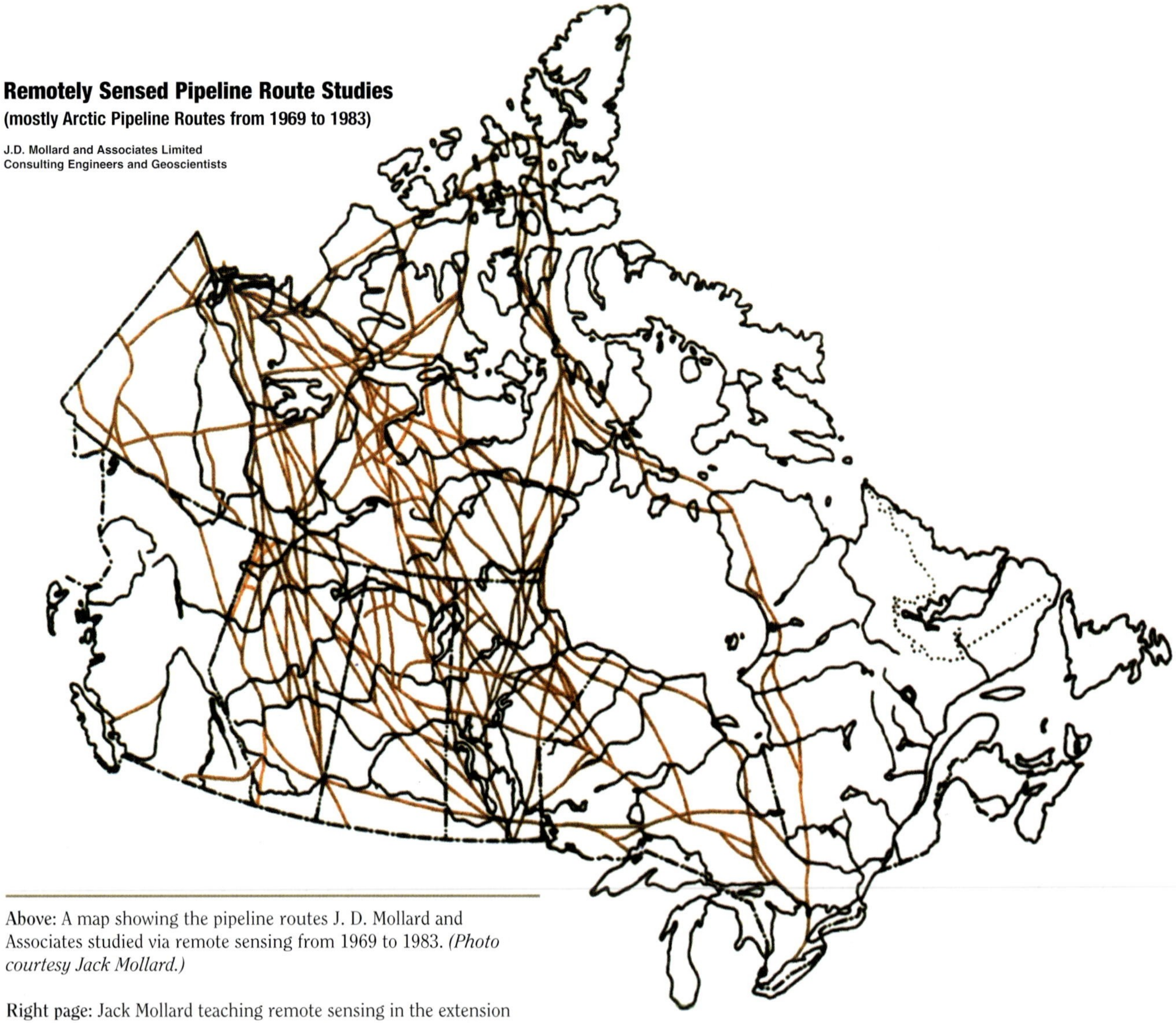

Above: A map showing the pipeline routes J. D. Mollard and Associates studied via remote sensing from 1969 to 1983. *(Photo courtesy Jack Mollard.)*

Right page: Jack Mollard teaching remote sensing in the extension department of the University of Alberta in 1967. *(University of Alberta Photographic Service.)*

photo landscape . . . for clues to sites to explore for groundwater. Vincent Beckie, hydrogeologist, and Jack Balzer, groundwater technologist, followed, supervising field drilling exploration and water-well pump testing to determine the safe well yield and water quality."

Searching for groundwater has been among the most challenging and rewarding tasks the company has undertaken, Mollard said. "That may be because I grew up on a combined wheat farm and purebred St. Bernard dog ranch outside Watrous, where we hauled water in barrels when it was 40-below-zero weather.

"In many of the Saskatchewan towns and villages where we discovered groundwater, the people there had no running water. They couldn't turn a water tap, flush a toilet, run a washing machine, or flood a skating rink.

"When you discovered a good drinking water supply for them, you'd think you'd found a gold mine."

CHAPTER 3 SOURCES

Publications

Annual Report of the Board of Highway Commissioners, 1912-1913.

Annual Report of the Department of Highways and Transportation, 1940-1941.

Annual Report of the Department of Highways and Transportation, 1957-1958.

Annual Report of the Department of Highways and Transportation, 1959-1960.

Cousins, Brian. "Transportation." *The Encyclopedia of Saskatchewan.* Regina, 2005. Canadian Plains Research Center.

Travacon Research Ltd. and Edmar Services Inc. *Saskatchewan Road/Rail Operation Project Evaluation, Final Report*. March, 1994.

Websites

Brandt Road Rail Corporation. Home page. February 24, 2006. http://roadrail.brandt.ca/

Personal interviews conducted by the author

Gentles, Tom. Former design engineer and deputy minister, Department of Highways and Transportation. November 16, 2005.

Mollard, Jack. President, J.D. Mollard and Associates. November 8, 2005.

Widger, Allan. Executive Director, Engineering Standards Branch, Department of Highways and Transportation. November 24, 2005.

Fibre optics networks transmit data digitally as pulses of light along hair-thin threads of glass. *(Photo courtesy Graham Bradley.)*

CHAPTER 4

Talk Amongst Yourselves . . . and to the World

"This modern convenience should be in every farmer's home just as soon as he could afford it . . ."

Above: An early Saskatchewan telephone exchange. Early Saskatchewan phones were hand-cranked magneto phones like this one.

Right page (left): Workmen install telephone lines on poles near an unidentified church in the 1930s.

Right page (right): Miss Emily Lander, Regina's first telephone operator.

(Photos courtesy SaskTel Archives.)

Building the Provincial Telephone Network

Saskatchewan isn't exactly crowded even now, but it's hard for us to imagine, a century on, just how sparsely populated it was when it became a province in 1905.

The "urban" population, living in four small cities (Regina, Saskatoon, Moose Jaw and Prince Albert) and about two dozen towns, was 34,909, with everyone else (and the total population of the Saskatchewan portion of the North-West Territories prior to 1905 was only 91,279) scattered in miniscule villages or individual homesteads. And this, as Premier William Martin pointed out a few years later, in a province "larger than any European country with the exception of Russia . . . twice as large as England, Ireland, Wales and Scotland combined . . . twice the size of Italy and twenty times the size of Belgium." (Love, p. 2.)

Rural people were isolated enough in the summer; in the winter, when travelling even a few miles to the nearest village could mean taking your life in your hands, more than one family must have felt they were the last people alive on Earth.

But coinciding with the birth of the province came an extraordinary flood of people. By 1914 the population exceeded 600,000, making Saskatchewan the most populous Prairie province.

The tide of immigrants eased the isolation, but didn't erase it. Although the cities swelled, homesteads remained widely scattered and linked by poor roads (where there were any roads at all). And the winters were still long.

It's not surprising, then, that when the Saskatchewan government decided to move into the telephone business, much of its focus was on providing service to rural areas.

The phone comes to town

By the autumn of 1882, just six years after Alexander Graham Bell patented his newfangled device for sending voice messages over wires, Regina already had 20 telephones in use on 10 private lines. All were owned and operated by Bell Canada. By December, the North West Mounted Police had its own system (eventually taken over by Bell). Bell set up its first Regina exchange in 1887 in Peter Lamont's bookstore, halfway between Hamilton and Rose Streets along what is now Saskatchewan Drive.

Other towns soon developed their own independent telephone services. A few even extended lines out to farmers (something Bell avoided).

Over the next few years, the incipient province's telephone networks grew. In 1904, Bell announced plans to upgrade the Regina system and build long-distance lines to Indian Head, Lumsden and Moose Jaw. By 1906, Regina telephones were linked to Winnipeg; Saskatchewan's first long-distance call was made between the offices of the *Regina Leader* and the *Manitoba Free Press* on November 6. (Long-distance calls cost 70 cents per minute during the day, half that at night.)

But about that same time there were growing complaints across the country about the near monopoly enjoyed by Bell Canada, and calls for public ownership of the telephone system. In 1907, Premier Walter Scott commissioned engineer Francis Dagger, the leading

Above: Operators work at the long-distance ticket distributing desk in the new Saskatchewan Government Telephones building on Lorne Street in Regina in 1955. When the long-distance toll network was modernized across Canada to allow direct-distance dial, Regina was one of only two regional toll centres, the other being Montreal. *(Saskatchewan Archives Board R-B5315-5.)*

telephone expert in Canada, to report on the state of the Saskatchewan telephone system and make recommendations for its future. Dagger's report, submitted on March 24, 1908, pointed out that Saskatchewan had 3,250 telephones, more than half of which belonged to Bell Canada—but that only 310 of them served rural people. And, he noted, of 121 municipalities, cities, towns and villages in the province, only 22 had local telephone systems.

Bell Canada had 166 miles of pole line, running from the Manitoba border to Regina to Lumsden. Even though the Moose Jaw-based Saskatchewan Telephone Company had more line—254 miles, radiating east to Regina, west to Mortlach, north to Tuxford and High Pound Lake, and southeast through Estevan to North Portal—only Bell provided long-distance service, via its link to Winnipeg.

Dagger recommended that the government should own and operate all long-distance equipment, while local exchanges should be left in the hands of municipal councils. But it was the rural telephone system that he said was of "supreme importance." To ensure efficient service and protect the subscribers from losing money, he said, rural municipalities should build and operate their own telephone systems under government supervision. (Love, p. 52.)

The government accepted Dagger's recommendations and passed legislation establishing the Department of Railways, Telegraphs and Telephones. James A. Calder was named commissioner and James Sutherland superintendent of telephones. A few months later, William Warren became the first chief engineer.

Rapid advances in technology

The most common telephones in the early years of the 20th century were oak-encased, wall-hung side-winders. Turning a bell-crank on one side spun a magneto, generating an electrical current that zipped along the phone lines either to another phone (if there was a direct connection) or to a switchboard in the central office, ringing a bell to signal an incoming call.

Each phone also contained a battery to provide the electricity that carried the voice signal. These batteries constantly caused trouble, since there was no way of knowing how long they would last and they had to be replaced by a trained repairman.

The early phone lines were made out of steel or galvanized iron (in some areas, calls were even sent along barbed wire). The wires would corrode, leading to loose and noisy connections. Nor were they very good electrical conductors. As a result, signal strength fell off quickly over fairly short distances.

The early wires were replaced by copper, which worked very well but cost four times as much—especially since the initial copper wires were as thick as a lead pencil. Immediately, engineers set to work to find some way to send more signals along thinner wire—in other words, to increase its bandwidth. One method was the phantom circuit, which allowed two pairs of wire to carry three voice circuits. (It's even possible to create "phantoms of phantoms" to obtain seven circuits over just four pairs of wire.) Better transmitters and better insulation also helped. Other innovations reduced interference and crosstalk, which plagued early systems.

One big improvement was the installation of a "central energy" system in the second storey of Bell Canada's Regina exchange building. A large battery in the building's basement provided power for both signalling and transmitting messages, eliminating the need for the bell-crank and the untrustworthy local batteries.

But the biggest technological advance didn't come from Bell: it was the 1907 installation in Saskatoon by the local Northwestern Telephone Company of the first automatic service in the country—which meant the people of Saskatoon were the first Canadians in history to be able to dial a telephone (see sidebar).

"The operation of the 'phone is simple"

On February 2, 1907, Saskatoon's *Daily Phoenix* newspaper explained in simple terms how one used the newfangled rotary telephone made possible by Canada's first automatic switch:

First of all, the receiver must be removed from the hook. To call, say No. 138, insert a finger in the hole in the dial opposite the first numeral, swing the dial to the left until the finger touches a point indicated, a finger-stop it is called. Remove the finger and let the dial return of its own accord, and repeat by placing the finger the next time in the hole opposite to the figure 3 and the third time opposite the 8. Connection is then complete with the number desired and by pressing the small button the bell on the other end rings and may be kept ringing as long as desired. The receiver should be placed to the ear as soon as the last numeral has been called. If there is a buzzing sound then hang up the receiver for that is the "line's busy" signal of the automatic 'phone. A special busy signal is one of the cute contrivances of the new system. (Love, p. 81.)

Buying and building lines

On May 1, 1909, the Telephones Department purchased the Bell system, consisting of 13 exchanges, 20 toll offices and 234 pole-miles of long-distance circuits radiating from Regina. On July 1, it purchased the Saskatchewan Telephone Company, which netted it 32 toll offices, 258 pole-miles of line and four exchanges, the largest of which was Moose Jaw. On October 1 it bought the Wapella-Harris Telephone Company. (Love, p. 89.)

As 1910 began, the provincial system consisted of 20 exchanges and 100 toll offices, serving 5,710 subscribers (3,412 local, 2,118 rural, and 180 municipal). (SaskTel.) The Department of Telephones set out to harmonize all of the existing equipment. Exchanges were remodelled, a new telephone central was built in Moose Jaw and additional wires were strung along the main government-built lines.

By February of 1909, 21 rural telephone companies had been organized, their incorporation paid for and their construction materials supplied by the provincial government.

In 1910 minister of agriculture W.R. Motherwell said in a speech in Saskatoon that "this modern convenience should be in every farmer's home just as soon as he could afford it, for nothing in recent years has so tended to popularize and modernize rural life as the introduction of the rural telephone." (Love, p. 92.) A farmer's telephone linked him to the outside world, provided a vital social connection with far-flung neighbours, and ensured that help—especially medical help—could be summoned if needed.

By February of 1909, 21 rural telephone companies had been organized, their incorporation paid for and their construction materials supplied by the provincial government. (Construction itself was often a challenge. For example, many companies had trouble keeping telephone poles perpendicular in marshy ground. Even a slight tilt could snap the wires.)

Most rural telephone companies installed party lines, averaging one subscriber for every pole-mile constructed, and limited to 10 subscribers per line. The party line, as *The Canadian Engineer* (the national engineering journal) pointed out in 1901, was an inexpensive, effective way to build telephone connections among a small number of rural customers. Yet it could also be used to increase connectivity among hundreds of customers over a very large area just by installing a small switchboard at a central location. (Love, p. 97.)

The party line relied on the old magneto technology, which required every telephone to have its own battery. Each subscriber was assigned a distinctive pattern of short and long rings. (The ring could be generated either with the hand-crank on the phone of the person placing the call or by a central operator.) A general announcement to all subscribers could be issued with one sustained ring.

Left page: As part of a massive modernization effort, engineers test new equipment in the new Saskatchewan Government Telephones facility on Lorne Street in Regina in 1955. *(Saskatchewan Archives Board R-A19021.)*

Left: Robert McKindoe installs telephone lines at Langenburg in July of 1937. *(Photo courtesy SaskTel Archives.)*

Above: A pile of hand-cranked magneto telephones discarded in favour of modern dial telephones provides a photo opportunity in Wapella in 1964. *(Saskatchewan Archives Board R-A19021.)*

Phenomenal growth

The construction of rural telephone lines proceeded at a phenomenal pace. By 1913, there were 337 farmers' companies, with 7,554 pole-miles of line and 8,024 subscribers. The government, meanwhile, had completed 2,679 pole-miles of trunk lines (in addition to the 492 purchased from Bell and other companies). The ratio of telephones to residents dropped from one for every 92 in 1908 to one in every 23 five years later.

By 1924, when the Department of Telephones issued an annual report that looked back on the development of telephony since 1908, 67,744 pole-miles of line had been erected and a quarter of a million miles of wire had been strung. There were nearly 100,000 telephones in service, and more than 64,000 belonged to rural subscribers—more than twice the number in Alberta and Manitoba combined. The ratio of telephones to people had dropped to one phone for every eight people.

The quality of construction within the Saskatchewan telephone system, the report said, was as good as anywhere in the world, and better than most. Cedar poles, full metallic circuits and standardized buildings made it possible for Saskatchewan residents to enjoy clear long-distance telephony right across Canada and into the United States as far east as New York and as far south as Kentucky.

The report noted that more than 500 pupils had attended the Saskatchewan Government Telephone School, which trained farmers to diagnose and repair lines and equipment. The Telephones Department's engineering staff stood ready to help with more complex problems.

The report concluded that Saskatchewan was "... today covered in great part with small systems, standardized construction to ensure service, operated as a convenience, managed locally, with a central government staff acting in an advisory and supervisory capacity, and all linked together with trunk lines erected and maintained, the whole giving a system operated by the Government and the people to meet the need of the various sections of the Province ... The Province has telephone service." (Love, pp. 112-113.)

Inquiries about the Saskatchewan system came from Europe, South Africa, Argentina, Jamaica, New Zealand and Australia, not to mention other provinces. Saskatchewan had become a world leader in telecommunications engineering.

It wouldn't be the last time.

STOCK PHOTO

“If the guys way up there in Gopher Flats can make this thing work . . .”

The SaskTel Fibre Optic Network

In the 1980s, SaskTel made a technological leap over the heads of telecommunications companies many times its size, with the installation of what was then the world’s longest fibre optic network: 3268 km in all, serving 52 communities.

But the impetus for that engineering marvel was, as for many other major Saskatchewan projects, politics.

“The government had decided that SaskTel was also going to do the cable TV operation in a large number of communities,” recalled Pieter Van Vliet, who was chief engineer of the Crown corporation from 1975 to 1980. “We were asked to go into the cable TV business, which was completely new to us. I collected a staff of about eight people and we set out to find out all there was to know about cable TV engineering.”

At the time, cable TV only existed in Regina, Saskatoon, Estevan and Weyburn. The signals arrived via microwave. SaskTel had a lot of analog microwave systems in place, but they didn’t have enough bandwidth to provide TV service in addition to the telephone signals they already carried. “We had to rethink how we would serve those communities,” said Van Vliet.

To make economic sense, the new system had to integrate video delivery with traditional telephony. And, since the government wanted the 52 smaller communities served by the system to be able to send video signals back along the system as well as receive them, it had to be bi-directional.

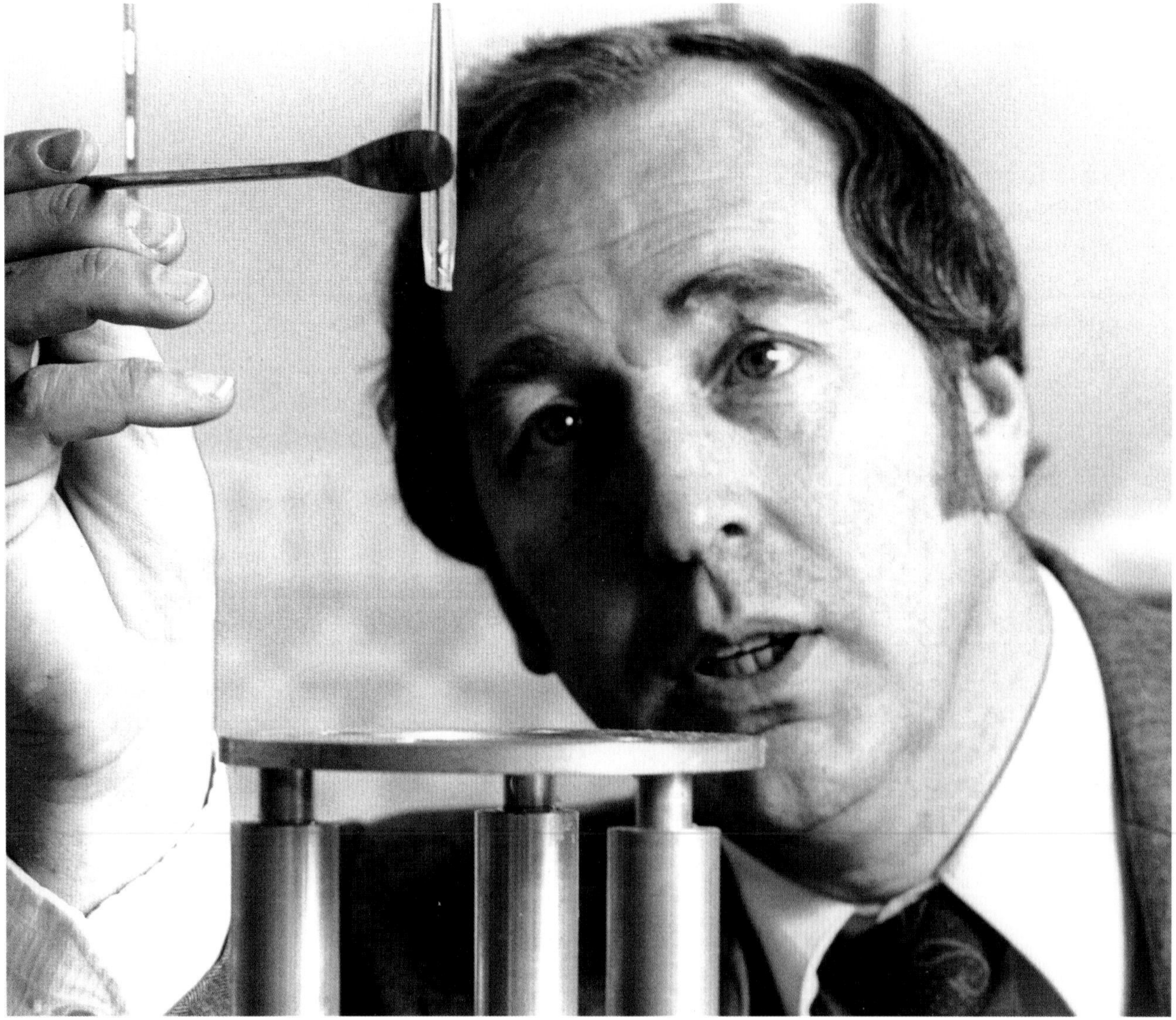

The new technology also needed to tie into SaskTel's efforts to modernize its entire service, which had been under way since the 1950s, back when SaskTel was still known as Saskatchewan Government Telephones. Countless hand-cranked magneto telephones had been replaced by modern dial telephones. Old pole lines had been replaced by buried cable. The long-distance toll network had been modernized to allow direct-distance dial (a Canada-wide effort in which Saskatchewan played a leading role: Regina was one of only two regional toll centres in Canada, the other being Montreal). And electro-mechanical switches were being replaced with computerized digital switching equipment.

By the late 1970s, SaskTel had ambitious plans to convert all of its switching stations to digital by the year 2000 (a goal actually achieved by 1995). But to do so, SaskTel needed a digital, broadband, bi-directional network—the same kind of network it needed to fulfill the government's cable TV goals.

Left page: S.R. Woodland, division general manager for the optical systems division of Northern Telecom Canada in the 1980s, examines the droplet formed at the end of a treated glass rod that is being melted into several kilometres of hair-thin optical fibre. *(Saskatchewan Archives Board R-B12844.)*

Above (top): SaskTel equipment digs a trench for fibre optic cable. *(Photo courtesy SaskTel archives.)*

Above (bottom): Fibre optic cable unspools into a trench as the SaskTel Broadband Network is installed in the 1980s. *(Photo courtesy SaskTel archives.)*

Right: A SaskTel team lays fibre optic cable alongside a railroad track in rural Saskatchewan in the 1980s. *(Photo courtesy Graham Bradley.)*

("Digital" meant information would be transmitted as a series of 1's and 0's, the language of computers. "Broadband" meant it would be able to send the vast quantities of data necessary for video. "Bi-directional" meant data could travel both ways along the network.)

Which technology?

Some details of the proposed new network had inadvertently been determined much earlier. In Saskatchewan, as anyone who has ever driven its highways knows, small communities pop up every 16 km or so along the rail network. This isn't by chance: that distance was a trade-off between how far steam engines could go without water and how far horses could haul a wagon full of grain. To serve smaller communities, the new network therefore needed to provide bi-directional access at approximately 16-km intervals.

Above: Fibre optics networks transmit data digitally as pulses of light along hair-thin threads of glass. *(Photo courtesy Graham Bradley.)*

SaskTel's engineers considered, and rejected, several technologies. Analog coaxial cable wouldn't do: its signal vanished in noise over long distances. Analog microwave wouldn't work: besides the lack of bandwidth in the microwave network, it wouldn't be economical to provide access to a microwave network every 16 km. Digital coax was suggested (Manitoba's telephone company even installed some), but it required frequent repeaters to boost the signal, and many enclosures to house those repeaters, making it too expensive. Digital microwave had the same access problems as analog microwave. Satellite distribution of TV signals was on the horizon, but bi-directional satellite communication requires expensive uplinks.

SaskTel needed something new to build what would be the first digital television network of such a length anywhere in the world. The engineers turned to fibre optics, which had seen only limited use in military applications, but appeared to hold great promise.

In a fibre optic system, data is sent digitally as pulses of laser light along hair-thin cables made of glass so ultra-pure that if the ocean were made of it, you could see the bottom, even at the deepest point. At the receiving end, computers translate digital data into an analog signal that generates a TV image or vibrates the speaker of a telephone.

Even today fibre optics sounds a bit science fictional. In the 1970s, it was really "out there." Former SaskTel vice-president Frank Degenstein, for one, was incredulous when engineer Bob Allen began talking about fibre optics. Whoever heard of transmitting telephone messages, much less television, over glass hairs?

But early experiments by SaskTel researchers convinced the corporation's decision-makers that fibre optics could meet the government's requirements, and so in 1980 SaskTel decided to push ahead with the Broadband Network (BBN).

A tender moment

The biggest problem was the cost of the cable, but the major suppliers suggested the cost could come down if the order was big enough. To which SaskTel replied (in the form of a tender), "Okay, how much will the cost come down if we order, say, 3200 km?"

Part of the tender package, recalled

Lawrence Fisher, the lead engineer on the project at the time, was a book, *The Atlas of Saskatchewan*. "We said, 'All the environmental conditions are in there, you've got to meet them.' It's interesting, we got more people wanting a copy of the book than were interested in the tender. I think we actually ended up generating some tourism for the province. People looked at the book and said, 'Gee, that's kind of an interesting place,' and then they came here."

In the end, Northern Telecom won the tender, and proceeded to build a factory in Saskatoon to manufacture the necessary fibre and other components.

The tender, recalled Pieter Van Vliet, produced "great astonishment," especially among the "Bell people, who thought if anything was going to happen with fibre optics it would be in Bell territory."

Graham Bradley was project manager for the first major component of the BBN, a 200-km stretch between Regina and Yorkton. "We had people come and say we couldn't do it," he said. "Bell came down and said it wouldn't work; we'd never hook that much fibre together. Little did they know!"

One of the main questions at the beginning—and one of the major contributors to cost—was how often repeaters to boost the signal should be placed. "If you could stretch the repeater spacing between, say, a toll centre and a community dial centre, you gained quite a bit," Van Vliet said.

"We had a unique way of bidding the thing," Bradley said. "If they had more repeaters . . . they were actually penalized in money, so the people with the best technology were going to win, no matter what they bid. We forced them to push their envelope."

Despite that, there was an excellent working relationship between the SaskTel engineers and those from Northern Telecom, Bradley recalled. "It was more like a technical bonding. It wasn't your typical supplier/customer/client relationship. We were all buddies. We'd been working together for a year before we started the damn thing. Everybody knew everybody, we'd all kind of worked together . . . it wasn't the money, it was the challenge of doing something new, which made a big difference. It went a lot smoother that way."

"Okay, how much will the cost come down if we order, say, 3200 km?"

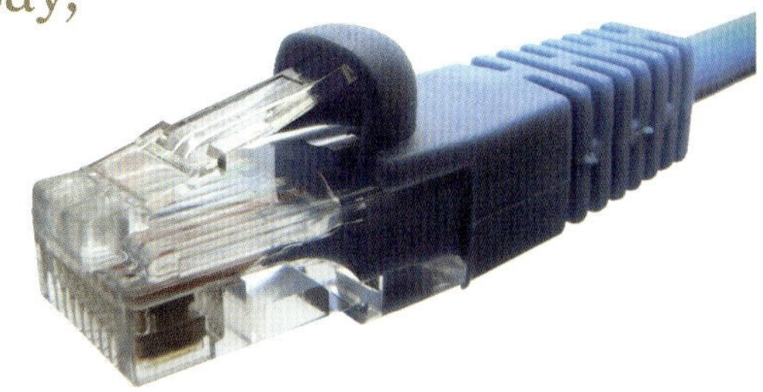

The distance between repeaters stretched from about 6 km to begin with, to 10 to 12 km, and then, after a change in the wavelength of the laser being used, 30 to 50 km.

Great gobbling gophers!

Not all of the network's innovative features involved high-tech components. One potential problem was averted early on: gnawing gophers (pocket gophers, not the more common Richardson's ground squirrel).

"In the States they actually had gnawed the whole cable—they chewed right through," Bradley said. "And with fibre optics, at least in those days, you could only splice it so many times because there's extra loss in the fibre every time you connect it together, so if it gets spliced enough times, the fibre's useless.

"We thought at 5 feet there'd be no gophers. Well, it turns out that at 5 feet there's even more gopher activity, 'cause that's below the frost line. So we armoured it (with steel tape). And everybody's armoured it ever since. Our contribution to the world!"

Saskatchewan in the Chunnel

In 1991 Van Vliet, who by then had left SaskTel and set up his own consulting firm, was called on to head up a technical audit of the French-English Channel Tunnel (Chunnel) project. (His two-man team included a SaskTel engineer and a quantity surveyor from a British company.) "We discovered a whole lot of things that needed to be fixed," he said. "We were also astonished by what we found out about the cost."

Two years later, the Chunnel's fibre optic data transmission network had been installed—but systems tests were failing because there were too many problems down in the tunnel. The project faced a fixed deadline (December of 1993), and if the fibre optic network didn't work, 11 other systems that depended on it wouldn't work either—in fact, pretty much nothing would work.

Van Vliet's solution was to bring in a "raft" of SaskTel techs with experience on digital carrier systems and sometimes on fibre optics. "A lot of these people had been in the districts, they were used to working on their own—if there was a problem they could solve it. They ended up going so much faster than both the French and the British that there was a lot of rivalry. We had to split the Canadian crews over the French and the British so they all had more or less equal levels of expertise. From then on we made great progress."

Many deficiencies had to be corrected before the system could be tested. The French had been using three crews. The first would go in and attempt to test the equipment in one of the 72 pairs of equipment rooms located between the running tunnels and the service tunnels. If the test failed, they'd move on to the next equipment room, while calling for a second crew to come diagnose the problem. Once the second crew made a diagnosis, a third crew went out to actually fix the problem. Since it could take three or four hours to get to a particular work station, and another three or four to get out again, a trip to a single equipment room could eat up an entire day.

By contrast, said Van Vliet, "The Canadian crews would go in, find out they couldn't do the test, diagnose, fix, then go back and do the test, all on the same visit."

Many other lessons from SaskTel's experience were applied to the Channel Tunnel project, Van Vliet said. "I assembled a team of people that I trusted, people that were experts in the planning and execution, who had a bulldog quality of not letting go of a problem. We set them in motion and gave them all the support. In the end it was terrific fun."

And yet, he recalled wryly, "We were never allowed to be part of the official ceremony when it was finally put in service. They made sure all the colonials were sent home by then."

Above (left to right): Dave Weick of SaskTel International and Henri Golia and Mike Mathews of Transmanche Link (TML) pose in late 1992 in front of one of the cutting heads used to excavate the Chunnel, now on display at the tourist information centre located near the tunnel railyard complex at Folkestone, England. *(Photo courtesy SaskTel.)*

Above: One innovation Saskatchewan provided the telecommunications world was the armouring of fibre optic cables against the depredations of pocket gophers, who liked to chew on them. *(Photo courtesy Graham Bradley.)*

The spin-offs

The BBN was completed in 1984, and for several years (until a slightly longer network was completed in Japan), Saskatchewan boasted the longest fibre optic network in the world—and more real-world experience with fibre optics than anyone else.

Bradley wrote a number of articles for telecommunication industry periodicals about the project. "That really opened the door," Van Vliet recalled. People from all over the world—the United States, Europe, Australia, New Zealand—all wanted to know how SaskTel had pulled it off.

"They thought, if a bunch of guys way up there in Gopher Flats can make this thing work, there must be something to it," is how Fisher put it.

The international interest led to the formation of SaskTel International, which became a separate entity on January 1, 1986. "It was set up to market our skill and expertise and engineering in the fibre optic network," said Van Vliet.

SaskTel's Research and Development (R&D) Group, which worked behind the scenes on the fibre optic network, also gained official recognition in 1986. It went on to pioneer the Saskatchewan computer networks Telidon and Agritex, forerunners of the World Wide Web, and was the magnet that led to the formation of TRLabs branches at the Universities of Saskatchewan and Regina. The R&D Group also did the initial work on digital video and video-on-demand, which contributed to the innovative MAX service now offered by SaskTel.

What Saskatchewan engineers learned through the construction of the BBN in the 1980s, and disseminated throughout the industry, contributed to the development of the Internet. "Without the essentially zero cost of bandwidth that comes with fibre, you can't afford Internet," Fisher said. "Without fibre, there would not be an Internet."

Since the completion of the BBN, SaskTel has forged ahead with many other projects based on fibre optics, and continues to be a world leader in the field. Saskatchewan still has more fibre optic lines per capita than anywhere else in the world.

As a result, small towns in Saskatchewan have better telecommunications service than some of North America's major cities: an engineering legacy to be celebrated.

CHAPTER 4 SOURCES

Books

Love, Ronald S. *Dreaming Big: A History of SaskTel*. Regina, 2003. SaskTel.

Websites

Saskatchewan Industry and Resources. *25 Saskatchewan Science Achievements*. "SaskTel Installs World's Longest Fibre-Optics Network." 2005. (February 10, 2006.) <http://www.sk25.ca/Default.aspx?DN=89,88,16,1,Documents>

SaskTel. *SaskTel - History*. (February 10, 2006.) <http://www.sasktel.com/about-us/company-information/history/index.html>

Personal interviews conducted by the author

Bradley, Graham. Project manager, first leg of SaskTel Broadband Network. Former head of SaskTel Research and Development Group. June 2, 2005.

Fisher, Lawrence. Lead engineer, SaskTel Broadband Network. June 3, 2005.

Van Vliet, Pieter. Former chief engineer of SaskTel. June 2, 2005.

CHAPTER 5

Black Gold, Saskatchewan Tea

An EnCana pumpjack near Weyburn, silhouetted against the prairie sunset. *(Photo courtesy EnCana Corporation.)*

"An exciting, intellectually challenging, and competitive period to be a practising geologist."

Above: Drilling Shell Big Muddy No. 1. *(Saskatchewan Archives Board R-A3207-1.)*

Wildcatting in Saskatchewan

At least three oil wells were drilled in Saskatchewan prior to 1900. Two of them came up dry; the third (drilled by the Canadian Pacific Railway at Belle Plaine) reported a "small gas flow." Other wells, just as unsuccessful, followed. Saskatchewan, it seemed, was not prime country for oil exploration. The "wildcatters" didn't give up, though. ("Wildcatter" is a nickname for oilmen who drill wells in order to find a new pool.) A well drilled at Kamsack in 1937 produced natural gas, but "unfortunately, the Kamsack gas wells provided only enough methane to fuel a farmhouse, and if it was of particularly good deliverability, the farmer's barn as well." (Halabura.)

Imperial explores

Imperial Oil, which opened a refinery in Regina in 1916 using oil from Wyoming, began exploring in Saskatchewan in the early 1900s. Over the next few decades Imperial's geologists systematically worked their way across the West, locating promising prospects and following them up with test drills . . . with, in Saskatchewan, no luck.

Nevertheless, more refineries opened in the province, including the Hi-Way Refineries Ltd. plant in Saskatoon in 1932, the British American refinery in Moose Jaw in 1934, and the Consumers' Co-op Refinery in Regina in 1935.

Then came the Second World War, and a new incentive to find oil and gas. In 1939, the provincial government offered huge exploration reservations to the Northwest Company (which became Norcanols Oil and Gas in 1942). Since it was a subsidiary of Imperial Oil, and Imperial Oil picked up its own large reservations in 1940, 1941 and 1942, Imperial effectively had access to most of the province.

Imperial placed Dr. J.C. "Cam" Sproule in charge of its Saskatchewan exploration and established a head office in Moose Jaw. Charlie Visser and Vern Hunter were placed in charge of the test drilling program, which totalled 15 wells from 1939 to 1946, eight labelled Imperial and seven Norcanols. They found traces of oil, but no hydrocarbons of commercial value.

In 1944, the new Co-operative Commonwealth Federation (CCF) government of T.C. Douglas was willing to extend the terms of the reservations to Imperial, but Imperial chose not to take the extension. That may have been because of its dislike of the Douglas government, or maybe just because it hadn't had any luck finding oil.

Over the next few years, other companies did find oil, mostly in the Lloydminster area, but the oil was heavy and thus of limited usefulness. The Saskatchewan oil industry was off, but, like heavy oil tipped from a jar in January, it wasn't exactly running.

Beginning in 1948, though, Saskatchewan entered what Steve Halabura, in his 1987 historical review of Saskatchewan wildcatting, termed "one of the greatest periods of exploratory activity that the Province has ever undergone." (Halabura.)

Boom!

One reason for this was that the CCF government began making available large exploration reservations at low cost. The 1945 regulations allowed subsurface permits of up to 34,650 acres (14 022 ha) and geophysical permits of up to 250,000 acres (101 171 ha). The permits lasted 21 months. Those who qualified could file for a permit, pay a nominal fee, and then go exploring for oil.

The other reason was the discovery of light crude oil at Leduc, Alberta in 1947. If light oil existed there, the oil industry reasoned, it must

"The most brilliant geologist to have worked Saskatchewan"

Tidewater's chief geologist was Dr. R.B. "Bob" Milner, a graduate of McGill University. Steve Halabura called him "perhaps the most brilliant geologist to have worked Saskatchewan . . . meticulous, curious and extremely imaginative." (Halabura, p. 12.)

H.M. Kirk, general manager of the Tidewater Group, said, "Without question, Bob Milner was the best geologist working in Saskatchewan in the 1950s. The Tidewater Group . . . hired some dozen or more young geologists just graduating from University. They arrived in Regina and Bob Milner would give them a six-week training in sedimentation followed by some 6,000 feet of well cuttings, and cores, for them to write up logs. This finished, they were sent out to sit on wells. They all came through good and moved to Alberta fields." (Halabura, p. 12.)

Above: An early oil well in the Lloydminster area. *(Saskatchewan Archives Board R-A17857.)*

Right page (left): Geologists study core samples at Stony Beach in May of 1952. *(Saskatchewan Archives Board R-A3207-1.)*

Right page (right): Setting up the derrick for Shell Big Muddy No. 1. *(Saskatchewan Archives Board R-A5438-8.)*

exist elsewhere on the Prairies. Geologists began looking for the underground formations that marked the Leduc field—specifically, Devonian reefs, ancient coral reefs more than 360 million years old, leftovers from a time when Saskatchewan was under the sea.

Explorers swarmed all over Saskatchewan. Lacking sound geological data (partly because the province had been scoured by glaciers, burying geological formations under a thick layer of glacial drift), obtaining permits was based on guesswork as much as science. The short duration of the permits caused many, faced with the choice of either conducting expensive exploration or losing their permits, to farm out their permits to larger companies.

Companies active during this time included Sohio (Standard Oil of Ohio), Tidewater Oil Co., Rio Bravo Oil Co. (the exploration arm of the Superior Oil Company), Phillips, Britalta Petroleums Ltd. and others. They conducted gravity surveys, measuring variations in the gravitational field to get a hint of the underlying geology (Sohio's crews took a reading at every section corner, covering the permit areas by jeep, horseback or snowshoe as required). They drilled. Tidewater flew aeromagnetic surveys, towing magnetometers behind aircraft to measure the intensity of the Earth's magnetic field. (Differences between actual measurements and theoretical values indicate the presence of different types of rock or different thicknesses of rock formations.)

Although many of the early wildcatters ignored southeast Saskatchewan, it was the first area the major companies moved into. Imperial and Shell began leasing freehold lands there, probably partly to avoid having to deal with the government in Regina. As well, a large company could deal with thousands of individual freehold leases more easily than the smaller wildcatters could.

Jack Porter, a geologist with Rio Bravo, was one of the first to recognize the potential of the formation in southeast Saskatchewan called the Mississippian subcrop. Deep tests drilled in the Souris Valley by Souris Valley Oils Ltd. in 1949 had established that a thick layer of Mississippian rocks found in the south of both Saskatchewan and Manitoba pinched out, due to erosion, further north. To geologists like Porter, this hinted there might be a large

stratigraphic trap, oil trapped within a reservoir because of an interior barrier, in that area. The discovery of the Midale oil field in 1953 bore this out, and launched the oil industry in the Southeast in earnest.

In the Southwest, a well drilled at Fox Valley in 1946 by a former securities salesman named Tom Keyes contained natural gas and opened the province's largest gas field, Hatton-Horsham. Keyes chose the exploration permits he wanted through the reconnaissance work of George Paul D'Aigneaux, who "claimed to be able to detect oil and gas at depth using certain black box methods of measuring radioactive emanations." (Halabura, p. 14.) D'Aigneaux was dismissed as a "doodle-bugger" and "quack," but he outlined the Hatton-Horsham-Fox Valley and Dollard areas as being possible sources of oil and gas—and, indeed, they've become major producing areas.

The oil industry in western Saskatchewan got off to a roaring start with strikes at Buffalo Coulee/Coleville in August of 1951 and Fosterton in 1952. A burst of exploration followed.

Today Saskatchewan produces more than 20 per cent of total Canadian oil production, making it the second largest oil producer after Alberta. About 24 per cent of Saskatchewan's oil is used within the province, with the remainder exported to the United States and other parts of Canada. Some 12,000 people are employed directly in the oil industry. Many more are involved in refining, upgrading and natural gas activity.

Although exploration continues, it's estimated that between 80 and 85 per cent of Saskatchewan's oil potential has been discovered. In future, therefore, the industry must try to get more oil out of existing pools.

Fortunately, engineers and geoscientists have come up with several ways to do just that.

A worker prepares to leave an EnCana pumpjack near Weyburn. *(Photo courtesy EnCana Corporation.)*

Horizontal Drilling in Saskatchewan

Who Knows Horizontal Better than a Prairie-Dweller?

In the 1980s, Malcolm Wilson, now Program Director for CO_2 Management at the Office of Energy and Environment at the University of Regina, was at Saskatchewan Energy and Mines (as it was then called) when Sceptre Resources came in and talked about setting up a demonstration horizontal well at Tangleflags North, located north and east of Lloydminster.

The reservoir there, thicker than most in Saskatchewan's heavy oil area, has water at the bottom and a small gas cap at the top. In 1987 Sceptre drilled a horizontal well low in the reservoir (but well above the water), then injected steam into the gas cap, theorizing that the steam in the gas cap would heat the oil and push it down into the horizontal well.

"The really important part of that," said Wilson, "aside from the fact it was a roaring success, is that it really demonstrated that in a fairly hostile environment a relatively small company could drill a horizontal well successfully, and you didn't have to be Imperial Oil to use this kind of technology."

That's not to say there weren't problems. Through human error, some equipment was lost down the hole, and initially the wrong equipment was used. Nevertheless, eventually the well was drilled, with about 400 m of horizontal section. Even though the horizontal part of the well was in unconsolidated sand—basically beach sand impregnated with heavy oil—it stayed open long enough for Sceptre to drill it out, then insert a slotted liner to keep it open.

"That particular location became an enormous success in terms of the amount of oil recovered," Wilson said, putting it at "60 per cent or so of the oil in place."

Sceptre went on to drill an equally successful well in the Southeast, where the solid rock meant that usually no liner was required. But a third, non-Sceptre project really launched the horizontal drilling boom in Saskatchewan, putting it in the forefront of horizontal drilling efforts around the world.

C.S. Resources and SaskOil co-operated on the project in the Winter oil field, described by Wilson as "one of those fields where there was a lot of oil, but again very, very viscous and over water, so average recoveries on primary were 1 per cent or less before you watered out." (A well "waters out" when it begins producing nothing but water. Water moves through rock or sand more easily than oil—especially when the oil is fairly heavy—so once water is drawn into a well, that's all that well will produce.)

The horizontal wells drilled in the Winter field demonstrated significantly improved primary recovery. This led to a mini-boom in land sales in the region, because areas where vertical wells no longer produced oil became viable again.

"Those very early wells were great successes," Wilson said. "They also demonstrated that in several different environments you could drill horizontal wells, and they weren't really prohibitively expensive to drill. So the end result of that was . . . when we hit the '90s, we basically went into a drilling boom driven by horizontal wells. At one point, in world figures for horizontal drilling, the United States was number one and Saskatchewan was number two. We were actually well ahead, in the early '90s, of Alberta."

> "It really demonstrated that . . . you didn't have to be Imperial Oil to use this kind of technology."

Saskatchewan horizontal wells were also being drilled in a much more varied set of environments than in the United States, where most horizontal drilling was in a single geological environment in Texas.

Today, horizontal drilling is commonplace in oil-producing regions around the world.

Innovative drilling techniques

PanCanadian Petroleum (now EnCana Corporation) in Weyburn started off with single horizontal wells, but now has wells with a single source on the surface that spread out in four directions underground, each horizontal segment 1000 m long. "PanCanadian really does deserve a lot of credit for being an extremely innovative company in the way it was drilling," Wilson said.

As another example, he described the company's extensive use of "underbalanced" drilling. When wells are drilled, a fluid (commonly called "mud," although there are all kinds of drilling fluids) is pumped into the hole to cool and lubricate the drill bit, bring the

Above: Hay bales and pumpjacks are both part of the scenery in the Weyburn oil field. *(Photo courtesy Petroleum Technology Research Centre.)*

Right page: A crew installs pipeline to a new well being drilled in the Weyburn oil field. *(Photo courtesy EnCana.)*

rock fragments from the drilling to the surface, prevent the hole from caving in—and keep pressurized oil, gas and water from entering the well prematurely, which could lead to a dangerous blowout. Since the Weyburn field had been in production for a long time, and no overpressure zones were expected, PanCanadian used lighter-weight drilling fluids than normal so that the fluid wouldn't infiltrate the reservoir. That in turn meant that the fluid wouldn't have to be cleaned up before any oil could be produced.

More innovation occurred in the Southeast as companies began experimenting with a novel way of getting more oil out of their fields by injecting something else into them.

The Weyburn Oil Field CO_2 Miscible Flood Project

In the 1980s PanCanadian Petroleum and Shell conducted the first pilot tests in the Weyburn and Midale fields of a new method of enhanced oil recovery, called carbon dioxide (CO_2) injection.

"Conventional wisdom would have told you in the '80s that you would not want to inject CO_2 into a fractured reservoir, that CO_2 would run along the fractures and not contact much of the oil," said Malcolm Wilson. "Shell demonstrated in the Midale field that that wasn't the case."

The success of the early tests led to larger pilot projects, and eventually to full-scale use of the process: EnCana has been injecting CO_2 into the Weyburn oil field since 2000. Apache Canada Ltd. began injecting CO_2 into its Midale unit in 2005.

As they pump in CO_2, they're pumping new life into the aging fields.

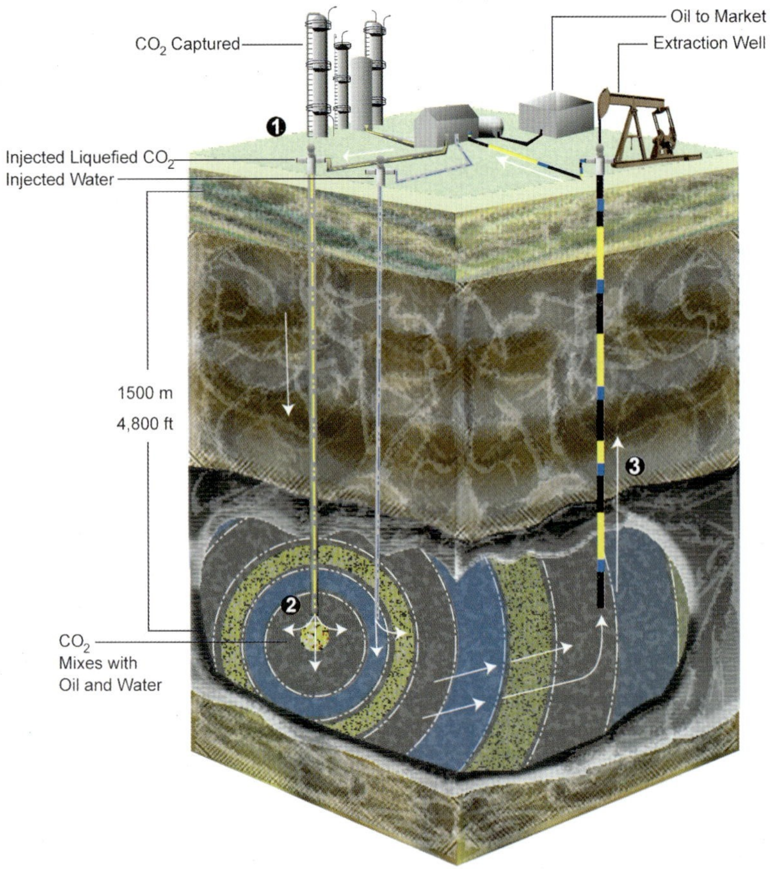

Above: How CO_2 miscible flood enhanced oil recovery works.

Right page: This diagram shows the area of the Williston Basin around Weyburn that was the focus of a recent four-year international study, led by the Petroleum Technology Research Centre (PTRC). The study, under the auspices of the International Energy Agency, was the first large-scale study in the world on the geological storage of CO_2 in a partially depleted oil field.

(Diagrams courtesy PTRC.)

Oil plays hard to get

The 180-sq.-km Weyburn oil field, discovered in 1954, is estimated to have originally contained 1.4 billion barrels of oil, making it one of Canada's largest medium gravity, sour crude oil reservoirs.

Only a small portion can be retrieved simply by drilling a hole. An oil well's peak production usually lasts only one to three months, then tapers off. You can keep drilling new wells up to a point, but eventually production from the entire field declines.

Once that happens, you have to turn to enhanced oil recovery (EOR). The first EOR step (taken in Weyburn in 1964) is usually water injection (known as waterflooding), which is just what it sounds like: a flood of water is injected into the bottom of the reservoir to increase its pressure and push the oil to the extraction wells.

How well waterflooding works depends on the field's geology. In the Weyburn field, the oil is retained in two layers of rock: the Marly zone, a chalky dolomite, and, underneath it, the Vuggy zone, highly fractured (and thus highly permeable) limestone. Waterflooding did a fine job of flushing oil from the Vuggy zone, but didn't make much headway in the Marly zone.

CO_2 miscible flood technology (CO_2 injection's proper name) involves injecting highly pressurized carbon dioxide into the reservoir. It's called "miscible flood" because the CO_2 actually mixes with the oil, swelling it and forcing it out of the small holes (pores) in the rock in which it is trapped. The process is aided by the fact that the CO_2/oil mixture is less viscous than oil by itself, allowing it to flow more readily to the producing wells.

Impressive and surprising

The results can be impressive—and surprising. David Hassan leads the group in charge of development of the Weyburn field for EnCana. "The reservoir is just extremely, extremely complex," he says. "It's very compartmentalized. It surprised us in terms of the type of response we've seen with CO_2 injection. Some areas responded

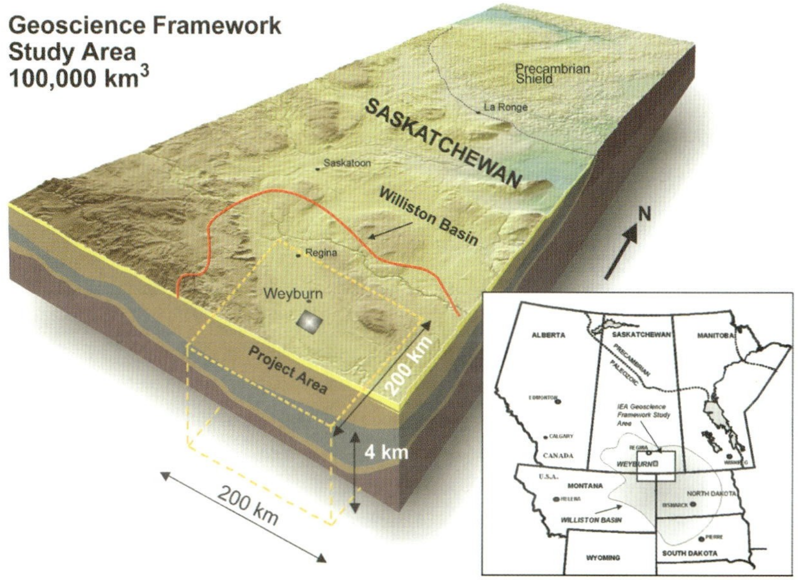

much more quickly than we had designed for, and some areas have responded much more slowly than we had expected."

The biggest surprise came when they started drilling horizontal infill wells in the EOR area in October of 2003, Hassan said. "One of the first three wells we drilled out there was the most productive well that's ever been drilled in the unit in over 50 years. The initial test rates were 2,000 barrels per day." In fact, he said, the well (nicknamed "Precious") was the fourth most productive well drilled in Western Canada in the last 10 years, based on the initial three months production.

Precious was just the best of eight or 10 astonishingly productive wells, Hassan said. Precious produced 300 cu. m per day, but "we've drilled a handful of wells that have had initial production over 100 cubic metres a day (600 barrels a day) . . . That has really been the surprise to us." In fact, Hassan said, "Weyburn has actually drilled three of the top 18 wells (excluding thermal oil sands/heavy oil) in the last two years, so the results continue to be impressive!"

The technology isn't new—there have been CO_2 injection projects in Texas for many years—but the success in Weyburn has been due to the way that geologists, geophysicists and engineers have worked together to understand the complex underground structure of the reservoir, Hassan said.

STOCK PHOTO

What's that smell?

The carbon dioxide used in Weyburn (and Midale) comes from the Great Plains Synfuels Plant, located near Beulah, North Dakota, which makes synthetic natural gas from coal.

The Beulah plant uses coal with sulphur in it, and as a result, the gas the plant produces—and the CO_2 coming north—contains methyl mercaptan, an extremely smelly substance. That meant that when the CO_2 injection plant in Weyburn came online, "Even the smallest little leak, because mercaptans are so odiferous, raised a big stink, literally," Hassan said. And tiny leaks were common, because pressurized carbon dioxide is a solvent for virtually any kind of rubber or plastic.

EnCana had a high-tech and a low-tech solution. The high-tech: different sealing materials and very careful assembly to minimize leaks. The low-tech: taping polyethylene bags over the fittings to concentrate the smelly gas, helping workers more quickly locate leaks.

One group of touring VIPs asked if the valves were new "because they were still in the bag."

Above (left): A pipeline delivers carbon dioxide to the Weyburn oil field for injection into the oil reservoir. *(Photo courtesy PTRC.)*

Above (right): A pumpjack at the Weyburn oil production facility. *(Photo courtesy EnCana.)*

Right page: The oil and agriculture industries mingle in the Weyburn oil field. *(Photo courtesy PTRC.)*

EnCana anticipates the annual average production level of the Weyburn field will reach 30,000 barrels of oil per day by 2008, compared to just 10,000 barrels of oil per day if the CO_2 miscible flood technology hadn't been implemented.

A geologist for every hole

Every hole drilled in the search for oil is drilled where it is because a geologist decided that was the place for it. But the geologist's work goes beyond just choosing the location. In his 1981 book *Roughnecks & Wildcatters,* Allan Anderson provided some anonymous quotes from geologists about what working in the oil industry was like:

"A geologist is on call at a rig 24 hours a day. There is no particular time when you are required to work. You can work any time you want, but normally you try to get some sleep when everybody else is sleeping. I would say your day begins around six o'clock. You're busy from that point until somewhere between 8 and 10 o'clock preparing a report to call in, on either a mobile or a land phone, to the particular oil company that you're contracted to. And after you've reported on the previous 24 hours' activities and what the rocks were and what was of interest in them, then you can heave a little sigh of relief for an hour or two. You get back, finally, to examining samples throughout the rest of the day and keeping on top of what's happening at the drill, or down at the bottom of the hole. You go on working till probably midnight, or whatever ... So you might say you work on the order of 18 hours a day, or at least you're fairly close to the operation for that period of time."

"When you hit oil, as opposed to gas or water, you can see it in the mud when it gets to the surface. You'll see streaks of dark oil just like streamlines. This doesn't happen too often, by the way, in the career of a geologist. There is only once in my career that I've actually seen oil. It did, indeed, excite me." (Anderson, pp. 76-79.)

Sequestering a greenhouse gas

All the CO_2 being used for CO_2 miscible flood EOR in Saskatchewan, because it's made from a fuel gas, would otherwise have been released into the atmosphere. Now it's being sequestered deep underground, at a rate of 5000 tonnes per day in the Weyburn field alone. In total, about 30 million tonnes of this greenhouse gas will be injected into the Weyburn reservoir over the life of the CO_2 project; another 8.75 million tonnes will be sequestered in the Midale reservoir.

A four-year international study at Weyburn, conducted under the auspices of the International Energy Agency, concluded there is a 95 per cent confidence that at most 0.2 per cent of that CO_2 will find its way into the atmosphere in the next 5,000 years. The study was the first large-scale study in the world on the geological storage of CO_2 in a partially depleted oil field. Now a new study is under way to see how the techniques and systems used in Weyburn can be used around the world. The hope is that

Above: A drilling rig hand on the floor of a drilling rig in the Weyburn oil field. *(Photo courtesy EnCana.)*

Right page: The Weyburn CO_2 injection gas compressors gleam in the Saskatchewan sunlight. *(Photo courtesy EnCana.)*

Weyburn will become the textbook example of how such things should be done, and that a kind of "rule book" can be created to help other oil companies follow suit.

JIVE talkin'

Another form of EOR being studied in Saskatchewan is JIVE (joint implementation of vapour extraction), also known as solvent vapour extraction, or "vapex." This involves injecting a gaseous hydrocarbon solvent (CO_2, propane, butane or methane) into a heavy oil reservoir. The solvent dissolves into the oil, making it less viscous and allowing it to drain into a horizontal well on a lower level for extraction.

Only 5 to 15 per cent of Western Canada's heavy oil reserves can be extracted without some kind of secondary process. The JIVE process could potentially increase the recovery rate to anywhere from 30 to 50 per cent—which translates into an additional 5 billion to 10 billion barrels of oil.

The SRC and EOR

The Saskatchewan Research Council (SRC) has played a role in all of the oil-recovery technologies discussed in this chapter.

The SRC was established in 1947, and began providing grants for applied research activities at the University of Saskatchewan. It expanded in the 1950s, getting its own building and staff, and identified key areas for long-term research, including water resources, fuels, transportation, metallic and industrial minerals, industrial methods improvement and the environment. In the 1970s, SRC began to earn more money from contracts for its services than it received in grants from the government.

Since its creation, SRC has written more than 2,600 reports in open literature and more than 2,700 confidential reports for commercial clients. Projects have ranged from mapping the groundwater resources of the province south of the Precambrian Shield to developing improved techniques for crop spraying to demonstrating the efficacy of satellite remote sensing for resource definition and management.

In 1986 SRC launched a multi-client project to determine if CO_2 injection was a good process for enhancing oil recovery in a light reservoir. It developed the modelling needed to calculate how much CO_2 was required and predict how the oil in the reservoir would react. SRC researchers also developed a set of pressure-volume-temperature (PVT) data needed for reservoir simulation, and (very importantly) determined the lowest pressure at which CO_2 would mix with oil, the "minimum miscibility pressure." SRC continues to work with EnCana by collecting PVT data.

Andrew "Sandy" Graham of EnCana said, "Without the work SRC did, we could not have gone to field. We needed the fundamental laboratory data to justify the capital expenditure of a commercial stage oil field operation."

SRC was also instrumental in the horizontal drilling boom. In the lab, SRC researchers designed a physical model to determine the viability of the horizontal well application. The lab work proved successful and led directly to the drilling of the first horizontal well at Tangleflags North.

SRC has also conducted (and continues to conduct) research on vapex technology.

Regina's NewGrade heavy oil upgrader, at night. *(Photo courtesy Saskatchewan Industry and Resources.)*

Upgrading Heavy Oil

Heavy oil has been described as "dark, thick, sticky and viscous" and "similar to blackstrap molasses." It's much harder to get out of the ground than light oil, and harder to do anything with once you have it. Which is why Saskatchewan became home in 1988 to Canada's first commercial heavy oil upgrader.

The NewGrade Energy Inc. upgrader, located at (and integrated with) the Consumers' Co-operative Refineries' light-crude oil refinery in Regina, breaks down heavy crude oil so that it can be refined into valuable petroleum products like gasoline and diesel fuel.

Before the NewGrade upgrader was built, Saskatchewan's heavy oil was shipped to the United States and used mainly to make asphalt. Demand and price fluctuated according to the time of year.

At full capacity, the NewGrade upgrader processes 8740 cu. m (55,000 barrels) of heavy oil a day. A second heavy oil upgrader was built in Lloydminster in 1992. The Husky Lloydminster upgrader processes about 10 800 cu. m (68,000 barrels) of heavy oil from both Alberta and Saskatchewan per day.

The upgrading process

Both plants use the same basic technology, although there are differences in detail. At the Husky Lloydminster upgrader, the procedure goes something like this.

Each barrel of heavy oil coming to the plant goes to the crude unit, where devices called desalters remove contaminants such as water, salt, clay and sand. The desalted crude is then heated in a tall vertical tank called a fractionation column. Lighter elements of the crude oil (about one-third of the barrel) are vaporized by the heating, rise inside the column, and condense and are drawn off at various levels, depending on their density. The lightest component of all is the diluent, added to the oil to make it easier to pump through the pipeline to the upgrader. The diluent is recovered from the top of the column and sent back to the production fields for reuse.

About two-thirds of the barrel of heavy oil remains at the bottom of the tower, looking very much like tar. It gets routed to the hydrocracker reaction unit, while the light fractions go to the hydrotreater.

In both of those devices, hydrogen (made by combining natural gas and steam) is added to the oil under high pressure. The hydrogen combines with sulphur in the oil to form hydrogen sulphide, which is absorbed in water and sent to a special plant that converts it into sulphur. The hydrogen also combines with nitrogen in the oil to form ammonia, which is also absorbed in water and extracted. The reactions take place in the presence of a catalyst, a substance that enhances chemical reactions but is not used up by them.

In the hydrocracker reaction unit, the hydrogen also reacts with metals still present in the oil and combines with the oil itself, increasing the ratio of hydrogen to carbon and converting approximately 60 per cent of the residue from the crude fractionation unit into lighter forms of hydrocarbon that can then be extracted.

The biggest difference between the ordinary hydrotreater and the hydrocracker reaction unit is the enormous pressure and heat involved in the latter—21 000 kPa (3,000 psi) and 425 degrees C. Designing equipment that can survive those kinds of pressures and temperatures is the biggest challenge in building an upgrader, according to Al MacLauchlan, operations manager at the Lloydminster facility. Among other things, it requires a reaction chamber with walls 28 cm thick.

Even after all that, some recalcitrant heavy oil remains. It's sent to the coker unit, where the last bits of lighter material are wrung out of it and sent to the hydrotreater. What's left over, mostly carbon, is sold as "petroleum coke," and used in the aluminum industry. What comes out of the hydrotreater is synthetic crude, ready to be refined.

Left page: An aerial view of the Husky heavy oil upgrader in Lloydminster.

Above: Another view of the Husky heavy oil upgrader in Lloydminster.

(Photos courtesy Husky Oil.)

The second biggest operational challenge in a heavy oil upgrader, said MacLauchlan (who has worked at both upgraders) is preventing coking where you don't want it. "Getting the conversion of heavy oil to lighter products tends to be a fairly sensitive process, and therefore you can easily get coking happening where you don't want it to happen, so you get fouling of some equipment."

To ebullate or not to ebullate

One technological difference between the Husky Lloydminster upgrader and the NewGrade upgrader is found in the catalytic reactors. Husky uses ebullated bed technology. "Ebullated" (which "is not a word in the English dictionary," MacLauchlan admitted) means that the catalyst, mixed with a liquid, circulates inside the reactor. While not unique to the Lloydminster upgrader, it is a lesser-known technology. Regina's upgrader uses a fixed bed system.

With a fixed bed system, the entire upgrader has to be shut down every year so the catalyst can be removed and replaced. The advantage of the ebullated bed technology used at the Husky Oil upgrader is that the catalyst, because it circulates, can be removed and replaced incrementally as needed, on a daily basis, allowing the upgrader to run indefinitely.

The importance of the heavy oil upgrader in the Lloydminster area is hard to overestimate. "It has stabilized the whole economy," MacLauchlan said. "Prior to the upgraders, with heavy oil in the Lloydminster area, if oil was up, everything was looking good; if prices were down, they were shutting down. Now they can produce a constant flow of heavy oil and a constant flow of synthetic light oil. The whole oil infrastructure in the Lloydminster area has become very stable. They've been able to develop the area."

The oil industry in Saskatchewan has come a long way from the early wildcatters, and has been marked by innovation after innovation. Research and development continue to meet the world's energy needs while minimizing the environmental impact . . . or, to put it another way, helping to build a safe and prosperous future.

CHAPTER 5 SOURCES

Books

Anderson, Allan. *Roughnecks & Wildcatters*. Toronto, 1981. Macmillan of Canada.

Scholarly articles

Halabura, S.P. (Steve). "New Oil from Old Ideas: An Historical Review of Wildcatting in 'Frontier Saskatchewan.'" Fifth International Williston Basin Symposium, 1987.

Websites

EnCana Corporation. Encana – Operations – Upstream. *Weyburn.* (March 6, 2006.) <http://www.encana.com/operations/upstream/ca_weyburn.html>

Foster Learning Inc. OTS Heavy Oil Science Centre. *Husky Oil Lloydminster Upgrader Launch Page.* (March 6, 2006.) <http://www.lloydminsterheavyoil.com/upgraderlaunch.htm>

Natural Resources Canada. "Notes for a Speech by the Honourable Ralph Goodale, Minister of Natural Resources Canada and Chair of the Cabinet Committee for the Economic Union, Saskatchewan Oil and Gas Show, Weyburn, June 9-10, 1999." (March 6, 2006.) <http://www.nrcan.gc.ca/media/archives/speeches/1999/199964_e.htm>

Saskatchewan Industry and Resources. *Oil in Saskatchewan.* (March 6, 2006.) <http://www.scs.sk.ca/vol/HTT/RR/SaskEnergyMines.htm>

Saskatchewan Industry and Resources. *25 Saskatchewan Science Achievements.* "Heavy Oil Upgrader the First in Canada." (March 6, 2006.) <http://www.sk25.ca/Default.aspx?DN=95,93,16,1,documents>

Saskatchewan Research Council. *History.* (March 6, 2006.) <http://www.src.sk.ca/html/about_src/src_history/index.cfm>

Other sources

Potter, Karen. Email correspondence. Corporate Relations Specialist, Saskatchewan Research Council.

Personal interviews conducted by the author

Hassan, David. Group Leader, Development, Weyburn Business Unit, EnCana Corporation. October 7, 2005.

MacLauchlan, Al. Operations Manager, Husky Lloydminster upgrader. November 5, 2005.

Wilson, Malcolm. Program Director for CO_2 Management at the Office of Energy and Environment at the University of Regina. November 7, 2005.

CHAPTER 6

Electrifying Developments

A Saskatchewan Power Corporation lineman at work near a farm in 1954. *(Saskatchewan Archives Board R-B2711.)*

"A goal that will tax our resources to the utmost."

Power to the People: Rural Electrification

Much of this book has to do with Saskatchewan "firsts," but sometimes, as the Bible noted, the first shall be last. So it was with rural electrification in Saskatchewan.

The Ontario legislature passed that province's *Rural Hydro Electric Distribution Act* in the early 1920s, but 25 years would pass before any of the Prairie provinces followed suit. Hard hit by both the Great Depression and a long drought, they had no money for electrification.

But in 1945, as the Second World War wound down, the Manitoba Power Commission set out to connect 35,000 farms within seven years. By 1953, 39,000 of the province's 50,000 farms had electricity. Meanwhile, in Alberta, 20,000 farms had been connected, even though Alberta farmers had to pay the full cost of power line installation, which averaged $1,100 per farm (Champ, p. 3).

In Saskatchewan prior to 1949, only about 1 per cent of farms had electricity. That contrasted sharply with urban areas: by 1926, 114 Saskatchewan communities had electricity, served up by individuals, small companies and municipalities. A number of larger private power companies appeared late in the decade.

Saskatchewan Power Commission is born

The Saskatchewan government got in on the act on January 18, 1929, legislating into existence the Saskatchewan Power Commission. The commission knew the task of bringing electricity to the province's farms would be formidable. As the Saskatchewan Reconstruction Council noted in its 1944 report, "The rural settlement in Saskatchewan is more scattered than in Manitoba, and in many areas farms are so large it might be impossible to give power coverage. It has been estimated that in Saskatchewan, with a saturation of 80 percent, there would be an average of about one customer per mile." (Champ, p. 4.) And unlike Ontario or Manitoba, Saskatchewan had only limited access to hydroelectric power, the cheapest form of power generation.

By the end of 1948, the commission operated 35 generating stations and more than 8800 km of transmission lines, but it had wired only 1,500 farms for power, mostly located close to the lines that linked towns and cities. In 1949 the provincial government passed *The Rural Electrification Act*, and turned the Saskatchewan Power Commission into the Saskatchewan Power Corporation (SPC).

Working from studies carried out over the previous few years, the government at first proceeded cautiously. SPC saw 1949 as a test year, and set a modest goal of 1,200 new farm customers. Saskatchewan's sparse rural population, which averaged just one farm customer per 1.6 km (SaskPower), meant a lot of expensive wire had to be strung on expensive poles by expensive workers to get electricity to a small number of customers who then might not use enough of it to make the operation economically viable.

How earth-return made it possible

Rural electrification might not have been possible at all, or at least not for many years, if not for a particular technical breakthrough: single-wire, earth-return transmission.

Electricity must flow in a circuit. The earliest rural lines in Saskatchewan used two wires: the current flowed out along one and back to the generator along the other. But between the wars, rural co-operatives in both Saskatchewan and the United States began experimenting with single-line systems. Instead of flowing back to the generator through a wire, the electricity made its way back through the ground.

In the United States, the single-wire, earth-return system didn't become very widespread because of fears it might be hazardous and/or interfere with communications. But during the late 1940s, the Saskatchewan Power Commission erected about 640 km of a single-wire, earth-return system for testing purposes. By 1949, W.B. Clipsham, the utility's chief electrical engineer, concluded that, thanks to the composition of Prairie soils, the one-wire, single-phase system would be both satisfactory and safe for use in Saskatchewan's rural electrification program. Communications interference could be avoided simply by placing power and telephone lines further apart. (White, p. 277.)

Using a single-wire instead of double-wire system is estimated to have saved SPC $125 per kilometre—which adds up.

Technical challenges

Robert G. Lawrence began working at SPC in 1963. Over 29 years he served first as vehicles and equipment engineer, later as chief

Left page: Holes for rural electrification power poles are dug near Parkbeg in May 1956. *(Saskatchewan Archives Board R-A13090-3.)*

Left: Power poles for rural electrification are loaded onto trailers at Parkbeg in May 1956. *(Saskatchewan Archives Board R-A13090-1.)*

The shocking truth

Electricity comes into a house at 220 V, then is split into two 110-V circuits with a ground in the centre. "If the ground is not good the voltage will drop on the leg that's got all the load on it," Lawrence said, "and the voltage on the other side will go way up. It would still add up to 220, but it could only be 60 on one side and the difference (160) on the other side. We had quite a bit of trouble with that in a few places, especially during the dry years where the ground resistance changed significantly."

This was not good for small appliances, but it could usually be solved by putting in a larger ground grid and driving ground rods down to where there was still a fair amount of moisture, or adding a neutral conductor to the system to take the return back to the town substation.

engineer and eventually as senior vice president of operations. The heyday of rural electrification "was an era I kind of missed out on," he said, "but I certainly heard about it."

One technical problem posed by the single-wire, earth-return system was keeping ground resistance down to an acceptable level, Lawrence said. In rocky terrain, the system wouldn't work at all. "For instance, it won't work at La Ronge."

Drought could also cause problems, because the drier the ground becomes, the less effectively it conducts. "Keeping your ground return resistance at a proper level in a country like ours, which is arid, requires due diligence at all times," Lawrence said. "We had several ground test crews working throughout the summer months doing ground testing and taking the necessary corrective action."

Finding an affordable conductor posed another challenge. The final choice, steel instead of copper (with steel-copper combinations in some areas), was made possible by another technical innovation: adopting a distribution voltage of 13 800 V, compared to the standard (for the time) 6900 V. The higher voltage made up for the poorer conductivity of steel as compared to copper. It also meant a smaller conductor could be used, and run for greater distances from high-voltage supplies. (White, p. 277.) Steel's lighter weight and greater strength also made possible spans of around 107 m, which reduced the number of power poles required.

"Those guys strung it so tight that in winter it was like a fiddle string . . . to keep ground clearances up as much as possible," Lawrence recalled. Ground clearance was especially important in rural areas, where farm equipment had to pass underneath the lines.

Rural electrification takes off

After the "test year" of 1949, rural electrification slowly ramped up. In 1951, the target was 3,500 new farm customers. And then, in the 1952 provincial election campaign, the incumbent Co-operative Commonwealth Federation (CCF) administration of T.C. Douglas promised to increase electrical service from 4,000 to 40,000 farms—in just four years. Douglas's government was re-elected—which meant SPC had to make good on the promise.

Charles E. Smith, rural electrification superintendent, wrote at the time, *"Forty thousand farms in the next four years* is a goal that will tax our resources to the utmost and calls for the most extreme degree of organization and planning." (Champ, p. 8.)

Since there were about 12,000 farms already hooked up, simple math showed that 28,000 would have to be connected over the next four years, or 7,000 per year—double the number the corporation had targeted in its most ambitious year to date, 1952.

Lawrence emphasized that while many people think of rural electrification simply as providing single lines to farms, there was a lot more to it than that. "This was a major part of the program but of equal importance was supplying power to the towns and villages that number in the hundreds through the province."

Supplying all those farms, towns and villages required the design and construction of an entire transmission system (about 4500 km) of 69-kV lines, a sub-transmission system (16 000 km) operating at 25 kV, and the necessary substations (more than 100) to step the voltage down as required. (Towns and villages were supplied at either 4160 or 2400 V.) And, of course, the routes the lines would follow and the places where the substations would be constructed all had to be planned and arranged, and the public educated through a series of community meetings.

Despite the scale of the challenge, SPC fulfilled Douglas's promise. 1956 turned out to be the peak year of the program, with 7,800 new farms connected, bringing the total to 40,000—about 47 per cent of all farms in the province. Hundreds of non-farm rural customers, such as schools and community halls, also received power.

After that, the pace slowed. Through the 1960s new farms were added at a rate of about 2,000 per year. The total number of connected farms stabilized at about 60,000 in 1966.

"The piece of equipment that was the mainstay in those days . . . was the production digging machine," Lawrence remembered. "All they really did was go along and dig holes as fast as you could." As many as 100 poles a day could be erected.

The total cost of rural electrification through the 1950s was calculated at $47.5 million, with the farmers themselves contributing more than $25 million—an average cost of $500 per farmer.

That contribution from farmers meant that another important element in the success of the rural electrification program was convincing farmers to hook up. They had to be sold on the idea that they actually needed electricity on their farms. The province also needed to convince them to use more electricity once they had it. "In my time, early times out there, there really wasn't one electrical line that I'm aware of . . . that really paid its way. They really did not have sufficient load to pay their way for the cost of the installation and the cost to maintain it and service it, and of course the cost of power," Lawrence recalled.

Self-help

In some areas, farmers banded together to install poles. SPC would clear the power line right-of-way of brush and other obstructions and mark the location of each pole, and provide the necessary tools and poles. Farmers would then form a crew that would prepare the poles for wiring and place them as required. An SPC wiring crew would return later to string the conductors. The farm crew was paid a flat amount of $6.50 per pole. A farmer and son working together on a 15- to 20-man crew could earn almost what they paid to get electrical service, or enough to cover much of the cost of wiring their buildings. (Anderson, pp. 119-120.)

In 1954, about 200 farms were served with self-help. Over the next three years, nearly 4,000 farms were served with self-help, more than 17 per cent of the total electrified during that time. (Anderson, p. 125.)

Long, lightly loaded lines also posed a technical problem. SPC has two levels of service voltage: 120 V plus or minus 5 V, and 115 V plus or minus 5 V. "Two things happen," Lawrence explained. "There's a capacitance effect, so on a long, lightly loaded line the voltage will rise. Trying to hold the voltage down with no load is a problem. Then the reverse takes place, and when the load comes on, the losses start to increase significantly and the voltage takes a dive. Holding the voltage to customers at the end of the line within acceptable levels becomes a fairly major problem and something that has to be watched at all times."

One important step toward making sure farmers made full use of their electrical connection was ensuring that the farms, which were usually wired by contractors or the farmer himself, had adequate wiring. In 1951 SPC hired a wiring supervisor to meet with farm groups.
A handbook on farm wiring requirements was also prepared.

In March of 1952 the corporation organized a two-week short course in rural electrification through the Extension Division of the University of Saskatchewan. Forty farmers took the course in 1952, only two of whom had any experience with electricity; the rest were expecting hook-up soon. The next year the course was offered again to 40 students; another 40 had to be turned away. The course was offered again in 1954 and 1955. (Champ, p. 15.)

Farm women get the picture

The province promoted electricity use not just among farmers, but among their wives. SPC created a character named "Penny Powers" to that end.

The first Penny Powers, Lillian McConnell, travelled five days a week to fairs, schools and community halls across rural Saskatchewan. Eventually as many as four Penny Powers travelled from fair to fair over the summer, demonstrating the use of electrical kitchen gadgets and how to adapt to electric ranges recipes once cooked with a wood- or coal-burning stove.

Women seem to have grasped the advantages of electricity in the

STOCK PHOTO

"Bad news for the little pigs"

The sad fate of some Swift Current-area pigs illustrates how important it was that farms be properly wired.

Bob Lawrence explained there's an electrical term known as "step potential." A voltage gradient radiates out from an energized grounded object, decreasing in magnitude as it moves away from the source. That means that when a person or animal is standing nearby, there can be a voltage difference between where one foot (or pair of feet) touches the ground and where the other foot (or pair of feet) touches the ground—the recipe for an electric shock.

"The step potential for a pig gets fairly high because they're long and their feet are spread a fair distance apart," Lawrence said. "When farmers or people that live in the rural areas had their systems installed, they had to have their system well grounded . . . If the system isn't well grounded, you can end up with a voltage drop between (in this case) the watering bowl and the ground.

"Say your heater is not good, or you've got a bad plug or a frayed wire, and you've got a bit of a leak to this aluminum watering bowl—bad news for the little pigs."

Of course, Lawrence noted, pigs are smart animals. It didn't take the survivors long to learn their lesson. "They just refused to go to this watering bowl."

home faster than men grasped the advantages of electricity on the farm. In the mid-1950s Saskatchewan's Royal Commission on Agriculture and Rural Life found that the impact of rural electrification on the practice of agriculture, technologically and economically, had so far been minimal—but that the social impact had been significant. The commission estimated that 90 per cent of the electricity used on farms was used in the farm home.

One farm woman told *The Western Producer*, "I tried all the switches and then I carried the coal oil lamps down to the basement, singing a triumphant little chant about 'No more lamp glasses to clean, no more lamps to fill.'" (Champ, pp. 19-20.)

In a 2001 interview, Lillian McConnell told the same newspaper, "The first thing homemakers wanted was an iron. The second was a refrigerator." (Champ, p. 17.)

At the 1953 Saskatchewan Farm Electrification Conference, a panel of farm women came up with a list of household equipment they'd like to buy. It included a washing machine, iron, tea kettle, refrigerator, deep freeze, electric range, floor polisher, electric food mixer, dryer, sewing machine, electric fan, toaster, heating pad, clock, roaster oven, deep fryer, automatic coffee maker and dishwasher. (Champ, p. 23.)

Many of those items, of course, were already taken for granted by people living in towns and cities. There were hopes that rural electrification would stem the tide of people—especially young people—leaving rural areas. "With running water, water heating and farm lighting, the farm home can be every bit as modern as the city home," claimed a 1949 article in *The Western Producer*. "With the same installations in the farm community buildings—the rink, curling rink, community hall, churches and schools—can all be modernized so that the 'night life' organized by the community can be as attractive as the 'night life' offered in the city." (Champ, p. 24.)

A booklet from the electrical industry stated, "Yes! An electrified farm home will go a long way toward keeping your family together . . . we can guess that you or your neighbour has a son who is determined he is going to leave the farm. If it is because he wants modern living and envies city life, the strongest argument you can give him is an electrified home." (Champ, p. 24.)

Farm demand mushrooms

But if farmers were slower than their wives to see all the advantages of electricity, that situation didn't last long. Seed cleaners, shop tools, fertilizer mixers, seed treaters, irrigation equipment, ventilation systems, watering systems, milking

Above: The meter box, electric connecting cord and ground wire are installed on a power pole before it is raised into position on a farm near Parkbeg in May 1956. *(Saskatchewan Archives Board R-A13090-6.)*

Above: The present-day Boundary Dam control room. *(©[2005] by SaskPower.)*

Right page: An aerial view of the brand new coal-fired Boundary Dam Generating Station in July of 1959. *(Saskatchewan Archives Board R-B4758-4.)*

machines, clippers, sprayers, feed crushers, electric incubators and brooders . . . the list of useful farm equipment that could be powered by electricity was both longer and ultimately more power-intensive than that used in the house. As a result, the big challenge in the two decades following rural electrification switched from simply getting power to customers to getting enough power to customers.

Despite all the effort in the 1950s to boost power consumption by farmers, no one had foreseen just how much farmers would eventually take to electricity. "The system was planned for about 5 horsepower per farm," Lawrence said. "That allowed for two times the growth, but nobody foresaw all the electrical appliances and all the electrical uses that manifested themselves in the '60s and '70s."

That doesn't take anything away from the original achievement, though, said Lawrence. "These guys did a great job . . . Those were the times of working-really-hard and playing-really-hard people. That's what they were. Really tough. A lot of them were boys right off the farm. They attended a SaskPower-designed-and-conducted training course and from there went out to the field to work as many hours as required.

"They did a heck of a job. You'd have a hard time doing it today."

'Lectricity from Lignite

When Captain John Palliser led his famous expedition through Saskatchewan in 1857, the Native people told him about outcrops of coal along the Souris River near Roche Percée, a rock formation not far from where the community of Estevan would eventually take shape.

It's not surprising, then, that the lignite coal of southern Saskatchewan was probably the first of the province's mineral resources to be exploited by Europeans. Studies from 1880 on into the first two decades of the new century detailed the coal resources of that part of the province. Mining of the coal predated the studies: farmers in the Willow Bunch, Wood Mountain and Cypress-Shaunavon areas dug it out of pits, or even drove tunnels into hillsides. This continued for decades. The files of Saskatchewan Industry and Resources and the Geological Survey of Canada record more than

> "I believe this is the way you should make power, using coal."

400 mines, most of which served individual families or, at most, small communities. (Saskatchewan Industry and Resources, p. 66.)

The first commercial lignite production began near Roche Percée in 1880: a group of entrepreneurs began shipping fuel to Winnipeg on barges down the Souris and Assiniboine rivers. In 1890, continuous commercial coal mining commenced. Ever since, coal production in Saskatchewan has centred on the Souris coalfields and the city of Estevan.

In the early years the big coal mines were underground operations, but they were all gone

Above: Underground in a Saskatchewan lignite mine early in the 20th century. *(Saskatchewan Archives Board R-A940-2.)*

Right page: The giant Mr. Klimax Marion 7800 dragline in operation in the Estevan coal fields of Great West Coal Co. in 1958. *(Saskatchewan Archives Board R-B8409-2.)*

by the mid-1950s. They couldn't compete economically with strip mining, which began in 1927.

Coal was originally used for heating, then began to be used industrially. A briquette plant established in 1921 lasted only a couple of years, but in 1927 Western Dominion Collieries Ltd. bought it, modified it, and resold it. Since 1931 that plant and an upgraded successor, now operated by Luscar Ltd., have successfully produced up to 100 000 tonnes of barbecue briquettes per year.

From 1940 to 1958, provincial usage of coal remained steady. That began to change rapidly in 1959, the year Boundary Dam Generating Station started up.

Lignite electricity begins

Using coal to generate electricity wasn't a new idea, but historically lignite coal had been overlooked as a likely source of electrical power because it contains less energy than its harder anthracite cousin. The energy contained in coal is measured in British Thermal Units (BTUs) per pound (one BTU/lb. equals 0.556 kcal/kg). Anthracite averages around 15,000 BTU/lb. (8340 kcal/kg). The energy in lignite varies, but is generally less than half that of anthracite.

But with millions of tonnes of the stuff readily at hand, it's not surprising that the first lignite-fired power generating plant in Canada was built in Estevan in 1930. It generated 2750 kW and provided power for Estevan, Bienfait, Taylorton and Roche Percee, as well as several towns east of Estevan.

In 1946 the Saskatchewan Power Commission purchased the plant and renamed it the Estevan Generating Station. It was upgraded several times, eventually reaching a capacity of 70 000 kW in 1957.

In 1959 SaskPower took a major step in lignite-fuelled power generation with the commissioning of the $27 million Boundary Dam generating plant on the west bank of Long Creek, about 5 km south of Estevan. Today it is still the largest lignite-fired power plant in Canada.

Mike Zeleny is the current manager. At start-up the plant had two generating units, each producing 66 MW; today it has six units with a total generating capacity of 872 MW. In an average year Boundary Dam by itself produces 30 per cent of the province's electrical power requirements. "In 2004, a typical generating year, we produced about 6000 gigawatt-hours of electricity and burned almost 5 million tonnes of coal," Zeleny said. "We burn about 15 000 tonnes on an average day at full load."

" There are a lot of challenges with burning lignite coal."

How Boundary Dam operates

The coal is mined just a few kilometres from the plant. Large draglines with buckets capable of holding almost 70 cu. m of material at a time strip away the topsoil to expose the coal, Zeleny said. "There are three different coal seams in the area, varying . . . from just a couple of feet to 10 or 15 feet thick."

The coal is scooped out of the ground and dumped into a coal-hauler truck which can hold about 145 tonnes. The trucks continuously haul coal to the plant and dump it into the stockpile, which is constantly being pushed into the coal-crushing facility.

There are three different stages of crushing, Zeleny said. "Coal can enter the plant right out of the pit, so you can have chunks the size of a car. We have primary crushers

The Boundary Dam Generating Station as it appears today. *(©[2005] by SaskPower.)*

that deal with the big chunks, then we have secondary crushers that get it to about 1-inch size. That coal is then dropped into the coal bunkers that are above and beside the boilers."

From the bunkers, the coal drops into pulverizers which crush it to the consistency of fine flour. Large primary fans constantly blow air through the pulverizers, carrying the coal dust through pipes into the boiler furnaces. "Once the boiler is at a certain loading the combustion is self-sustaining," Zeleny said. The result is a constant fireball inside the furnaces with a temperature of around 1370 degrees C, and "sometimes a little hotter."

Surrounding the furnace are "hundreds and hundreds" of water-filled tubes 5 to 7.5 cm in diameter, depending on the size of the boiler. "The water in the tubes is heated very quickly, rises due to thermodynamics, and turns to steam," which continues to be heated, Zeleny said. "The steam is at roughly 1,000 degrees Fahrenheit (537 degrees C), and about 2,000 pounds per square inch

(13 790 kPa) when it leaves most of our boilers." This high-pressure, super-hot steam is directed onto the blades of the turbines, spinning them at 3,600 revolutions per minute (rpm). The turbines turn the generators, and the generators produce the electricity.

During his 29 years with SaskPower, Robert Lawrence managed Boundary Dam and oversaw the building of the Shand Power Station, Saskatchewan's newest lignite-fired power plant, also located near Estevan. (There's a third coal-fired plant, Poplar River, near Coronach.) "There are a lot of challenges with burning lignite coal," Lawrence said. "First of all, it has a very high moisture content, about 30 per cent. Just handling a product with that much moisture is a problem. In the early days they dealt with a lot of slagging in the boilers. It was not uncommon to open a boiler door and see slag running down the walls. If this happens it tends to plug your unit up and eventually shuts it down.

"A lot of ideas were tried. Actually taking a shotgun and shooting the slag off was tried. But what seemed to work the best was lime injection . . . to raise the melting point of this material." In the early days at Boundary Dam they injected "chicken grit" (calcium carbonate), he said. "We were the biggest users of chicken grit anywhere."

Boundary Dam now deals with the moisture in lignite coal in a very straightforward way, Zeleny said: air enters the pulverizers at a temperature of 370 degrees C. "(The water) is mostly evaporated, instantly—which absorbs part of the energy of that hot air, of course, but that's the way it's designed to work."

Burning coal produces more than just heat. "There's carbon monoxide, ash, carbon dioxide, a little bit of sulphur dioxide, nitrogen dioxide," Zeleny said. "The pollutants, so to speak."

About 10 per cent of the coal burned at Boundary Dam turns into ash. That translates into 1500 tonnes of ash per day. "We actually collect about 99.8 per cent of all particulate that comes out, so you won't see any ash coming out (of our smokestacks)," Zeleny said. "As a matter of fact, in summertime you

Environmentally friendly coal

Robert Lawrence said the 279-MW Shand Power Station, commissioned in 1992, was "basically my baby because it was built according to what I wanted. It's the first zero-discharge plant built in Canada."

The plant has a number of environmentally friendly features. For instance, powdered limestone and water are used to remove sulphur dioxide emissions. Other measures halved nitrogen oxide emissions. A closed-loop, zero-discharge water management system ensures water used in the plant only enters the environment through evaporation. A high-efficiency electrostatic precipitator collects 99 per cent of the fly ash.

Not only that, but adjacent to the power plant is a greenhouse. "The idea came from a meeting with people from the Tennessee Valley Authority who purchased a rainforest in Central America to offset CO_2 production," Lawrence said. "We came back, our planning people did the calculations, and came up with the million-tree idea: 1 million trees planted per year would offset the CO_2 production from the 300-megawatt unit."

Although SaskPower's board of directors expressed some skepticism when he first proposed the idea, Lawrence said, "I think it's been fairly successful. They kept up with planting trees to beat heck."

In 1993, Shand Power Station won a Power Plant Award from *Power* magazine in recognition of its advanced environmental design.

Today Saskatchewan gets about two-thirds of its power from lignite-fired power plants . . . and it's looking at building another one.

won't see anything." (In the winter emerging water vapour turns into fog on contact with the cold air, producing huge white plumes.)

About 10 per cent of the heavy ash is sold for mixing with cement. "The more we can sell, the better it is for us," Zeleny said. "We do deposit a considerable amount in our ash lagoons, which have been in service since 1960 or thereabouts. There's an ever-increasing amount of ash out there."

Everything on a larger scale

Today Saskatchewan gets about two-thirds of its power from lignite-fired power plants . . . and it's looking at building another one.

"I believe this is the way you should make power, using coal," Lawrence said. "Natural gas is far too expensive to use, and besides natural gas has a lot of other uses. Lignite is the way of the future and we should get on with it."

Right: An innovative feature of Saskatchewan's newest coal-fired power plant, the Shand Power Station, commissioned in 1992, is this greenhouse. Planting seedlings grown here helps to offset the carbon dioxide emissions from the plant. *(©[2005] by SaskPower.)*

Right page: The main street of Island Falls Village, which housed construction workers, in 1929. *(Saskatchewan Archives Board R-A959-1.)*

Island Falls

Building a power plant is a challenging engineering project wherever it's located, but that challenge is heightened when the project happens to be deep in Saskatchewan's northern wilderness.

Copper sulphide ore was discovered near Flin Flon, Manitoba, in 1915, and in the late '20s the Hudson Bay Mining and Smelting Company set out to mine it. To provide electrical power for the mine's development and operation, the company decided to build a hydroelectric plant. After extensive exploration, it settled on a site called Island Falls. (It got its name from three low waterfalls now submerged by the plant's reservoir.)

Island Falls is located on the Churchill River, which rises in western Saskatchewan and exits into Hudson Bay at Churchill, Manitoba. In its journey of 2120 km, it falls almost 400 m, with much of that descent concentrated in numerous falls and rapids—making it an ideal river for hydroelectric purposes.

"It's a really neat place."

Getting there is half the fun

The Island Falls site was 22 km west of the Manitoba-Saskatchewan border—and almost 100 km from the mine site. The only way to get to it was by 170 km of canoeing and portages. Before construction of the power plant could begin, roads had to be built between the site and the nearest railhead. Work began in the summer of 1928, and by the following winter about 70 km of forest road had been cleared and graded.

Camps were established along the route and connected by a telephone line. In the summer, equipment moved part of the way by water, so docks were built at the end of each lake, and several large scows were constructed. However, most of the heavy materials were brought in during the winter by road trains that consisted of a 100-hp tractor, six sleighs, and a heated caboose. Loads averaged 70 tonnes but could be as high as 110 tonnes. Each trip took between 36 and 40 hours. By the time construction

Above (left): The Island Falls hydroelectric plant as it appeared in 1930. *(Saskatchewan Archives Board R-B9756-1.)*

Above (right): From the beginning the Island Falls hydroelectric plant has employed local people. Here members of the original Native work crew from 1930 pose with their superintendent. *(Saskatchewan Archives Board R-A25014.)*

Right page: An aerial view of the Island Falls hydroelectric plant as it appears today. *(©[2005] by SaskPower.)*

was over, almost 32 000 tonnes of freight had been delivered to the construction site. About 70 per cent of it was hauled in during the first winter.

During the winter of 1928-29, permanent camp buildings were erected while two sawmills prepared lumber for construction.

The Churchill River Power Company, Ltd., a subsidiary of Hudson Bay Mining and Smelting, oversaw the project. The construction and engineering work was contracted to the Fraser-Brace engineering company of Montreal, which built the plant at a cost of $7.1 million. Concrete was used extensively because of the local availability of clean sand and gravel and because using local materials reduced the supplies (especially steel) that had to be hauled to the site.

Building begins

Plant construction began in May of 1929. There were two rapids at the site, split by an island. Two rock-filled cofferdams, one above the site and one below, diverted water from the south channel. Once excavation for the powerhouse and substructure, sluice sections and tailrace of the main dam was complete, and the concrete of the substructure had been poured, the north channel was blocked and the river diverted through the undersluice while work continued.

Meanwhile, Lang and Ross, Ltd., of Sault Ste. Marie strung steel-core aluminum cables on 356 galvanized steel towers to transmit power to the Flin Flon mine, 95 km away.

A power station for a power station

You can't run a mine without power, but ironically, you can't build a power station without power, either. The power needed for construction at Island Falls was provided by generators until a temporary power plant could be constructed at Spruce Falls on Duck Lake, a little over 20 km northeast. Construction of a wooden dam and powerhouse housing two 1,250-hp units began on October 4, 1928, and power began flowing to Island Falls on March 28, 1929. In 1931, with the Island Falls construction completed, the Spruce Falls plant was dismantled. Its generators, however, continue to provide power to Island Falls: they're now "house units" in the main plant.

The first generating unit in the station began providing power to Flin Flon on June 12, 1930. The concrete spillway dam was completed on August 23, 1930, and surplus water was released to cut a new channel to Sandy Bay. It re-entered the old river channel 3 km downstream and formed an island on which the town of Island Falls was later located.

Still going strong

More generating capacity was added as the Flin Flon mine grew. By 1959 Island Falls had seven generating units, producing 109 MW. All seven of those units continue to operate, according to Gerry Millette, Manager of Northern Hydro for the Saskatchewan Power Corporation, which assumed ownership of the plant in 1981 and took over its operation in 1985.

"We rebuilt the first three units in the mid-1990s and plan, over the next seven to eight years . . . to refurbish the next four units," Millette said, explaining that SPC upgraded the turbines on the oldest units, boosting them from about 12 MW to 14 MW. Some work was also done on the spillway dam's concrete, which is now deemed "to be good for another 30 years."

Up until 1967, Island Falls was not only a power plant but also a community of about 200 people, made up of the families of operators, electricians, machinists, administrators, labourers, and others employed to run and maintain the generating station, which was staffed 24 hours a day, seven days a week. The houses were quite modern, and heated—naturally—by electricity. But once technology advanced to the point where the plant could be run remotely from Regina, the community was abandoned. Today all that is left is a two-storey brick commissary that houses visiting staff as necessary.

The plant is staffed seven days a week, but only eight hours a day. When SaskPower first took over the plant, staff came from Nipawin every week, and only a few labourers were hired from the nearby village of Sandy Bay. In 1997, however, SaskPower signed an agreement with its unions and with the village of Sandy Bay, and launched a hiring and training program for the Aboriginal people in the area.

"Here we are eight years later and we've got employees from the village, Aboriginal employees, hired and trained," Millette said. "They're doing all aspects of the operation, so not very often will we send employees up from Nipawin any more. That's a big change."

Almost unchanged

Island Falls has changed little over three-quarters of a century. In the control room, Millette said, "mixed in with all the old dials are some electronic equipment now, so we're integrating that into the plant technology, but as far as the plant itself . . . it's not much different.

"Of all the plants we look after, it's my favourite," he said. "It's got a lot of nostalgia, a lot of history.

"It's a really neat place."

STOCK PHOTO

Interesting times

Hydroelectric plants need water. Too little is bad, but so is too much. In its 75 years, Island Falls has seen both extremes.

In 1941 the flow in the Churchill River dropped so low that the water shortage became critical. A control dam at Rocky Falls enabled some water storage in Reindeer Lake, but after that year's near crisis the Churchill River Power Company built a permanent regulating dam at Whitesand Rapids, 5 km below the lake outlet.

The opposite problem occurred in August and September of 2005. "We exceeded the previous high water, which was in 1997, by about 500 to 550 cubic metres per second," Millette said. "The tailrace elevation went up 17 feet from normal. That took us about a foot and a half over top of the tailrace deck, so we had to take steps to prevent water from entering the plant on the back side. We managed to keep it out and keep the plant operational." With the water that high, though, power production fell from 100 MW to just 64 (because the water had less distance to fall and therefore imparted less energy to the turbines).

"The water also washed over our plant access road by about 4 feet, something that had never happened before. We had to rebuild it," Millette said. "We had some interesting times."

The Centennial Wind Power Project

Saskatchewan is windy.

There's a lot of energy in moving air—and since the dawn of history, humans have harnessed that energy to move sailing ships, grind grain, pump water—and, in the last hundred years or so, to generate electricity.

Generating electricity from wind is simple in principle: as the wind blows, it turns a windmill or set of blades, which drive a generator. Hydroelectric plants work the same way—it's just that air is considerably less dense than water, and it thus takes more air than water, moving at the same speed, to carry the same amount of energy.

In the past two or three decades, with growing concern about the pollution produced by burning

Left: An aerial view of a small portion of the Centennial Wind Power Project near Swift Current. *(Photo courtesy SaskPower International.)*

"That's your fuel . . . so you want to go where it's most abundant."

fossil fuels, there has been a concerted effort to harness wind power on a scale sufficient to be of use to major power utilities—including SaskPower.

SaskPower's first foray into the use of wind power was twofold. First, it agreed to purchase electricity from the SunBridge Wind Power Project, a $20 million joint venture between Suncor Energy Inc. and Enbridge Inc. that operates 17 turbines generating 11.2 MW of electricity near Gull Lake. Second, it built its own Cypress Wind Power Facility, not far from the SunBridge one. The SunBridge project was up and running in 2002, and that same year the first nine turbines were installed in the Cypress facility. Seven more were added in 2003, bringing the total to 16, and the total generating capacity to 10.6 MW. That gave SaskPower a total of 21.8 MW of wind-generated power, enough to meet the needs of about 9,000 Saskatchewan homes.

In 2006, that total increased sevenfold, with the commissioning of SaskPower International's Centennial Wind Power Project: 83 turbines, more powerful than the ones at SunBridge or Cypress, capable of putting out a total of 150 MW—enough to power 64,000 homes.

In the beginning was the wind

SaskPower first began to seriously explore the viability of exploiting Saskatchewan's wind resource as a means to produce electricity in the early 1990s. A wind resource study undertaken in 1993-94 identified southwestern Saskatchewan as having the best conditions. The study involved the use of 10

30-m-tall wind monitoring towers placed around the region.

Mark Peters was project leader for both the Cypress Wind Facility and the Centennial Wind Power Project. "The collected data was used to produce a very rough, crude map of where the likeliest, windiest locations would be," he said. "If you look at that map, Gull Lake is identified as one of the hot spots." Suncor and Enbridge built their project near there, and, following some further study, SaskPower decided in 2001 to build the Cypress Wind Facility in the same area.

There was more to that decision than just identifying where the wind blew, although that's obviously important. Environmental constraints also had to be considered. "Are we right in the middle of a migratory bird path? Or are there some rare plants or animals in the area?" Peters said.

Constructability and site access are also important. "Is it really rough terrain? Is it going to be hard to bring in some very large components? Is it going to require building a lot of infrastructure?"

There are "stakeholder issues" that require discussions with local landowners and municipal, provincial and federal government officials.

Finally, a wind power plant needs to be somewhere that it can be readily connected into the power grid, and close enough to minimize "line loss," the loss of electricity due to the resistance inherent in even the very efficient conductors used for power transmission.

Left page (top): Workers prepare to pour the concrete foundation for one of the wind turbines at the Centennial Wind Power Project. *(Photo courtesy SaskPower International.)*

Left page (bottom): One of the rotors is held suspended by a crane so that workers can attach the nose cone to it. *(Photo courtesy SaskPower International.)*

Left: Workers attach a nose cone to one of the massive rotors used in the Centennial Wind Power Project, prior to it being hoisted into position. *(Photo courtesy SaskPower International.)*

Choosing a new site

In 2003, when choosing a location for the Centennial Wind Power Project, SaskPower decided to look at the province as a whole. Using sophisticated computer modelling that included topography, atmospheric modelling and vegetation, wind maps were drawn up for most of the settled portion of Saskatchewan, and combined with other information pertinent to developing a wind power facility. "We knew where our transmission system existed, we had some rough ideas as to which locations would be better than others to interconnect with respect to the grid, we also knew which areas were the hot spots in terms of wind," Peters said. Seven possible locations were identified, and wind monitoring towers were put up, "just to verify that what the computer models told us was indeed true," Peters said.

Right page: A rotor with 39-m-long blades is lifted into position for installation on the bus-sized nacelle atop one of the Centennial Wind Power Project units. *(Photo courtesy SaskPower International.)*

Ultimately a site near Rushlake Creek, about 25 km southeast of Swift Current, was selected and subsequently named the Centennial Wind Power Project.

The first problem was getting equipment and components to the site. The turbines and blades are manufactured in Denmark by a company called Vestas Wind Systems A/S.

The Cypress Wind Facility uses Vestas V47 turbines. The 47 refers to the total diameter, in metres, of the rotor: each of the three blades is 23 m long, and an additional metre is taken up by the hub. "A 23-metre-long load isn't that difficult to transport," Peters said.

But the Centennial Wind Project uses Vestas V80 turbines. Each blade

STOCK PHOTO

Construction challenges

Every major engineering project faces challenges. Some can be planned for. Others can't.

Take tornadoes, for instance.

"We had a tornado," Peters said, "the day we intended to lift the first turbine components.

"We had a crane ready to lift, we had a rotor assembled on the ground, we were going to lift the top power section, the nacelle and the rotor into place. That day they decided not to lift because there was bad weather approaching.

"That bad weather ended up being this tornado that . . . ended up hitting right where this crane was located. The crane toppled over, eight blades were damaged beyond repair, and (there was) other damage to equipment. Thanks to the hard work and dedication of the entire project team, we were still able to substantially complete the project in a single calendar year."

Peters isn't sure what damage a tornado would do to the finished turbines. "They're not designed to withstand tornado-force winds, but the site itself is spread out over a relatively large area, around 6 miles end to end, north to south around 3 miles . . . It wouldn't take out all of the turbines."

Anyway, Peters said, "Now that we've had one tornado, we're hoping we won't have to have another one."

is 39 m long. There are also very large and heavy nacelles, each about the size of a bus, that sit on top of each tower and house the gearbox, generator and other equipment.

"I don't know how difficult it was for Vestas to figure out the logistics of getting the turbine components to site," Peters said, "but what they ended up doing was shipping them from Denmark by boat to Houston, Texas, and then from Houston to the site on rail. As luck would have it, there's a rail spur that hasn't been used for I-don't-know-how-many years, maybe a decade, that just happened to be quite close to the project site. They actually delivered, by rail, these turbines right to the project site." The components then had to be trucked 10 to 15 km, depending on where in the roughly 30-sq.-km site they were needed.

Then there were the tower components. Each of the 83 turbines rests on top of a three-section steel tower 67 m tall. (The Cypress Wind Facility towers are "only" 50 m tall.)

The tower sections were manufactured by Hitachi in Saskatoon, then trucked to the site. There they were erected on concrete foundations—foundations which required the delivery of more than 3,500 loads of concrete to the site, also by truck.

That added up to a lot of heavy traffic. "Initially we thought, well, we're lucky, we have these three paved highways coming fairly close to the project area," Peters said. "But those three highways are fairly thin-membrane asphalt, and as soon as you start running heavy trucks on them, they tend to break up."

To minimize damage to the local infrastructure, the team identified several different routes to the site via gravel roads. The concrete trucks took one route and the tower components came via a different route, and both were separated to a certain extent from the roads on which the turbine components arrived from the rail line, "although there was some overlap there."

Putting it together

The other major pieces of equipment on site included large, heavy-lift-capacity cranes. To construct one unit, first the concrete foundation was formed, then the bottom tower section was bolted to it (using 144 anchor bolts). The second tower section was bolted on top of that, and the third on top of that. After that the nacelle was lifted into place and bolted on. The blades were bolted on to the hub on the ground to make a rotor, then the complete rotor was lifted into place and bolted on to the main shaft, which stuck out the front of the nacelle.

Each unit, from the ground to the top of the rotor when a blade is pointing straight up, is 107 m tall: taller than any building in either Saskatoon or Regina by several storeys.

Each turbine is capable of generating a maximum of 1.8 MW of power, but they don't reach full generating capacity until the wind is blowing at 15 m per second—54 km per hour (kph). However, they begin

to generate some electricity as soon as the wind speed passes 15 kph.

The rotor always turns at 17 revolutions per minute (rpm) when generating power, and it will operate at any wind speed up to 90 kph (although it never generates more than 1.8 MW). Each nacelle contains sensors that monitor wind speed, wind direction and temperature (among other things), and feeds that information to a built-in controller. The controller ensures that the nacelle is always turned into the wind. When the wind speed exceeds 90 kph, the controller puts the turbine into "survival mode," turning the hydraulically controlled blades edge-on into the wind so the turbine stops turning. It will also shut the turbine down if the temperature falls below -30 C. (Although, similar to a block heater on a car, the turbines have internal heating mechanisms to enable them to keep operating in less extreme cold.)

The turbines are designed to last 20 years under turbulent, gusty wind conditions; they may last even longer in Saskatchewan. Although winds coming out of the mountains are turbulent and gusty, Peters said, "by the time they get out to Saskatchewan they're fairly smooth. We have very good wind conditions from that point of view."

Turbine reliability has increased over the past few years. The expectation is that the Centennial Wind Power Facility turbines will be available to produce electricity "at least 95 per cent of the time." They require a certain amount of maintenance every six months or so, and more complete maintenance annually.

Matching wind power to the grid

One challenge for the personnel in the power grid control centre in Regina will be to adjust to the fluctuating nature of wind power. "As wind comes up and drops off, and the amount of megawatts we produce out of the wind plant goes up and down, you have to bring on additional sources of generation or ramp down other sources of generation. Depending on how quickly that happens, how quickly the winds pick up or drop off, you have to decide what the most cost-effective source of generation is that you're going to bring on. At a coal plant . . . you can't ramp up or ramp down very quickly, so if the wind is coming up quicker than what you can ramp down a coal plant, there's going to be a bit of a challenge there. But we've got other sources like gas or hydroelectric power that can complement in those situations."

The Centennial Wind Power Facility is just part of what SaskPower calls its "Green Power Portfolio." The goal is to ensure that all new electricity needs in Saskatchewan until 2010 are met with environmentally friendly power sources that don't add to greenhouse gas emissions.

But there's another good reason to harness the wind.

As Mark Peters put it, "That's your fuel—and it's free."

Left page: A worker oversees the mating of a rotor to the drive shaft of the turbine housed in the nacelle atop one of the units in the Centennial Wind Power Project. *(Photo courtesy SaskPower International.)*

Left: Turbine components, shipped from Houston, approach the Centennial Wind Power Project site aboard a freight train, using a spur line that had been all but abandoned until reactivated for the project. *(Photo courtesy SaskPower International.)*

CHAPTER 6 SOURCES

Books

Anderson, Dave. *To Get the Lights: A Memoir about Farm Electrification in Saskatchewan.* Victoria, 2005. Trafford Publishing.

Kupsch, W.O. & Hanson, S.D., editors. *Gold and Other Stories as told to Barry Richard.* Regina, 1986. Saskatchewan Mining Association.

White, Clinton O. *Power for a Province.* Regina, 1976. Canadian Plains Research Center.

Other publications

Champ, Joan. "Rural Electrification in Saskatchewan During the 1950s." Prepared for Saskatchewan Western Development Museum's *Winning the Prairie Gamble* 2005 Exhibit. http://olc.spsd.sk.ca/DE/saskatchewan100/researchpapers/WDM/Rural%20Electrification.pdf

Saskatchewan Industry and Resources. *Geology, and Mineral and Petroleum Resources of Saskatchewan 2003.* Miscellaneous Report 2003-7.

Websites

Rutherford, Dave. *Island Falls Saskatchewan: 1929-1967.* 2005. (Feb. 20, 2006.) http://www.islandfalls.ca/

Saskatchewan Industry and Resources. *Coal.* (February 20, 2006.) http://www.ir.gov.sk.ca/Default.aspx?DN=3549,3541,3538,3385,2936,Documents

SaskPower International. *Centennial Wind Power Project.* (Feb. 20, 2006.) http://www.saskpowerinternational.com/spi/invest/centennial/centennial.shtml

Personal interviews conducted by the author

Lawrence, Robert. Former chief engineer and former senior vice president of operations for SaskPower. December 9, 2005.

Millette, Gerry. Manager of Northern Hydro for SaskPower. December 1, 2005.

Peters, Mark. Project leader, Centennial Wind Power Project. November 14, 2005.

Zeleny, Mike. Manager, Boundary Dam Generating Station. December 9, 2005.

CHAPTER 7

Rock Stars

Miner working at the Dominique Janine underground uranium mine at Cluff Lake. After 22 years of operation, the decommissioning process at Cluff Lake began in 2004. *(Photo courtesy AREVA Resources Canada Inc.)*

“These rings permitted the development of potash mining in the province.”

The Blairmore Ring

Today, Saskatchewan is semi-arid. But from about 400 million to 375 million years ago, it was very, very wet, because most of it lay under water, beneath what is now called the Elk Point Sea.

For millions of those years, massive organic reefs in the area which now stretches from Great Slave Lake to the Yukon prevented the water in the Elk Point Sea from circulating freely into the open ocean to the northwest. That, combined with a hot, arid climate, meant that, just like the Dead Sea or the Great Salt Lake today, the Elk Point Sea became saturated with dissolved minerals, which precipitated out.

The result was multiple layers of minerals on the bottom of the sea: up to 215 m of salt (sodium chloride), calcium sulphate—and potash (potassium chloride). In southern Saskatchewan, there are four zones of potash near the top of the layers of the minerals formed from evaporation in the Elk Point Sea, buried at depths ranging from 900 m to 2750 m. (Vigrass.)

And a good thing, too, because today potash, used primarily in the manufacture of fertilizer, is the Saskatchewan economy's most important mineral resource. In 2004, sales of potash totalled $2.2 billion. This province is the world's largest producer and exporter of potash, accounting for fully one-third of global production and capacity.

But it might never have reached such heights if not for a singular engineering achievement known as the Blairmore Ring.

Where's the water coming from?

Potash was first discovered in Saskatchewan near the town of Radville in 1942. Its discoverers were looking for oil, not potash. But potash they found, at a depth of 2332 m. The next year saw a similar discovery near Ogema.

Western Potash Corporation Ltd. tried to set up a solution mine (look for more on solution mining later in this chapter) near Unity in 1951. That failed, so in 1952 the company sank a mining shaft. They had the dubious distinction of being the first to discover the most serious difficulty with mining potash in Saskatchewan: water. Water poured in in such quantities that the shaft had to be abandoned.

In 1956 the Potash Company of America drilled its first shaft at Patience Lake, near Saskatoon. Mining operations began in 1958—and ceased 11 months later, in 1959, due to flooding.

The problem, said Mark Fracchia, General Manager of the Potash Corporation of Saskatchewan's Lanigan mine, is that while Saskatchewan potash deposits are typically located about 1000 m below the surface, at 400 m there's a significant water-bearing formation called the Blairmore.

"Once you completed that shaft all the way down to the mine horizon at a 1000-metre depth, the water pressure coming from that particular formation . . . was pretty significant, in the 1,500 psi (pounds per square inch) type of range (measured at the mine level 1000 m below). The potential was there for significant inflow and then for not being able to develop the mine at all."

In fact, that's exactly what happened at Patience Lake: the shaft was closed from 1959 until 1965.

Rings of power

But early in the 1960s a technology was developed that could prevent the inflow of water from the Blairmore Formation: the Blairmore Ring. "The Blairmore Ring is a cast iron ring, the dimensions of which will vary from shaft to shaft . . . typically weighing up to about 28 tonnes," said Fracchia. "These things are stacked inside the shaft during construction to keep the water out."

Before you can install the rings, though, you have to hold back the water by other means. "Engineers came up with a system to freeze the water in the area that they were putting the shaft . . . by drilling down and inserting pipes to circulate chilled fluid," Fracchia explained.

Left page: Workers install a Blairmore Ring in a Saskatchewan potash mine. *(Photo courtesy PotashCorp.)*

Above (left): Potash mining at Rocanville. *(Photo courtesy the Saskatchewan Mining Association.)*

No foundries in Canada could cast such enormous pieces of iron.

Once the water-bearing formation was frozen, the rings were stacked in the shaft, above, through and below it. "There could be as many as 100 of these rings stacked one on top of the other. They were sealed against the shaft walls and also against each other with a grout or water-resistant chemical." With the rings in place, the water-bearing formation could be allowed to thaw.

The first Blairmore Rings, some up to 5 m in diameter, were installed in the International Minerals and Chemicals Corp. potash mine at Esterhazy. No foundries in Canada could cast such enormous pieces of iron, so various European foundries took on the task. (The most recent batch of Blairmore Rings were made in Ukraine for repairs to the shaft at the Potash Corporation of Saskatchewan's Allan mine in 1999-2000.)

Similar systems were already in use in Europe, specifically Germany, where most potash mine shafts have "tubbing" (rings). They're also used in Russia and anywhere else high-pressure water is encountered and a dry shaft is required, Fracchia said.

One of the Blairmore Rings stood in Rotary Park in Saskatoon from 1965 to 2001 as a tribute to the people who built the first potash mines in the area. In 2002, that ring was moved to the John G. Diefenbaker International Airport, where visitors can see it on arrival in the city.

It deserves its place of honour. As Fracchia put it, "It's safe to say these rings permitted the development of potash mining in the province."

Left: A crane installs a Blairmore Ring into its new location, on display at the Saskatoon airport. *(Photo courtesy PotashCorp.)*

Right page: Potash piled up in storage at PotashCorp's Rocanville mine. *(Photo courtesy PotashCorp.)*

The Patience Lake Mine Conversion

The Patience Lake mine that had been opened in 1958 and closed in 1959 due to flooding reopened in 1965. For the next 20 years it operated as a conventional mine . . . but the water wasn't done with it yet.

Mark Fracchia used to work at the Patience Lake mine. "Back in about the mid-1970s, as the mine was developing out toward the southeast corner of the mining area, an inflow developed . . . but it was kept under control by drilling and grouting. It just seeped for another 10 years."

Where did the water come from? "All the evidence pointed to the Blairmore again," said Fracchia.

"An increasing problem"

In January of 1986, miners noticed that "seeping" no longer described the problem. "The initial increase was rapid," Fracchia said. "Then it seemed to hold steady at several hundred gallons per minute. People

Above (left): In a conventional potash mine, borers dig out enormous rooms separated by thick pillars. *(Photo courtesy PotashCorp.)*

Above (right): One of the dredges at Patience Lake works in the settling pond, recovering the potash that has crystallized out of the cooling brine. *(Photo courtesy PotashCorp.)*

Right page: The Patience Lake potash mine as it appears today. *(Photo courtesy PotashCorp.)*

initially weren't that worried about it. Pump stations were installed, pumping excess water to surface . . . It wasn't until about mid-1986 that they realized they had an increasing problem on their hands."

If the incoming water had been brine (salt water), there might have been more of a chance to slow it. Brine in a potash mine is usually saturated, and so doesn't enlarge an opening it's flowing through by dissolving the potash (though it may by erosion). The Mosaic potash mine in Esterhazy, for instance, has an inflow of brine that can be controlled by injecting chemicals and grouts. A fresh-water inflow, on the other hand, quickly creates a larger opening by dissolving the potash it's pouring through.

By the time the leak became serious, the Patience Lake mine had been in production for more than 20 years, using what's called a "room-and-pillar" mining method. This involved cutting rooms about

6 m wide and 2.5 m high, separated by supporting pillars about 60 m long by 30 m wide. Some of the rooms became travelways or conveyor ways leading to and from the shaft. After 20 years, there were 13 or 14 such travelways leading to the portion of the mine where the water was pouring in.

Two scenarios studied

The first thought was that the flooded areas could be sealed off with concrete bulkheads, but even as the company commissioned the first bulkhead, it hedged its bets by hiring engineering firms to conduct studies on how to redevelop the mine in case the flooding proved intractable. Two scenarios were studied. One was creating new tunnels from the existing shafts, well above the potash beds, and ramping down to the potash level a safe distance away from the mine workings. The second was converting the mine into a solution mine.

Mine personnel working on the solution option began doing tests to determine how readily potash could be dissolved and the quantity and quality of potash that could be crystallized in a pond. Engineers created computer models to determine the probable production rate of a solution mine. "We brought in so-called experts," said Fracchia. "Unfortunately we soon realized there weren't any experts in that field. There were experts in solution mining, but not in converting a flooded mine to a solution mine."

The decision to abandon conventional mining at Patience Lake was made in February of 1987, by which time most of the engineering studies had been completed. The shafts were sealed above the mining level with concrete bulkheads and by July of 1987 the mine had filled with water.

"But the market was in a downturn at that point," said Fracchia, "so that delayed the decision for a little bit. Rather than make it a full conversion the company decided to develop what they called a pilot plant . . . and that was approved in September of 1987." The pilot plant started pumping brine from the recovery well to the pond in December 1987. In April 1988 brine pumping ceased. Next, the mill ran for three months to process product from the pond. The success of that operation led to the decision in August 1988 to proceed to a full-scale conversion.

Patience Lake had about 800 km of open mining rooms full of brine. In theory, the water entering those rooms had to dissolve only a few centimetres of the massive pillars to reach its saturation point—the point at which it couldn't dissolve any more potash. "The concept . . . was to drill wells at two different ends of the mine," Fracchia

Above: The Patience Lake potash mine at sunset, with a crystallization pond in the foreground. *(Photo courtesy PotashCorp.)*

explained. "For the full-scale conversion, we put in nine recovery wells . . . which we used to pump the liquid to the surface, and we drilled 10 injection wells to inject it back underground. Also, we carved out a 120-acre production pond (four 30-acre ponds in series) from an existing tailings brine pond area."

"That mining technique really didn't exist anywhere in the world."

A new way to mine

The Patience Lake mine conversion was the first of its kind in the world, so maximizing its efficiency required a lot of research and fine-tuning. "We used consultants to develop a computer simulation for solution mining, with some initial but limited success," Fracchia said. "We also developed computer simulations in-house for the mine and crystallization ponds."

One challenge was to determine exactly what was happening underground as the brine circulated. "We were trying to find the optimal path to inject this brine to make sure it would stay in the mine as long as possible to maximize its potash concentration," Fracchia said. They injected tracer chemicals into each of the wells, then analyzed the brine from the recovery wells to get an idea of how long it took brine to go from one well to another. "Then we would superimpose that information on a mine plan and try to determine where the brine actually moved, to optimize production and see where else we could put wells to mine more productively in the future."

The tracers were readily available chemicals: weak organic acids and amino acids. But by the time those tracers made it to recovery wells, they were diluted into the parts per trillion range. "The trick there was analyzing those at a very, very low concentration," said Fracchia. "The analysis was much more difficult than the actual injection."

Submersible pumps in the recovery wells lift nearly saturated brine to the surface, where it's piped to crystallization ponds. The outside air cools the brine, and since warm brine can hold more potash in solution than cold brine, potash crystallizes out. (This is one reason the plant is most productive in the winter, usually pumping brine only between about October and April. Another is that in the summer the higher evaporation rate results in potash and salt precipitating out together, which degrades the final product.)

"Potash settles to the bottom of the pond, and the remaining brine is heated and reinjected back underground," Fracchia said. "So it does a closed loop."

The potash is recovered from the crystallization pond by two floating dredges, each of which has a bucket wheel on the front that scoops the potash from the bottom of the pond into the dredge. The slurry is pumped to a station on the shore which in turn pumps it to the mill for processing. "One of the two dredges that's still used today is the original dredge that was leased for the pilot plant that was built back in 1987, albeit with some major modifications and improvements," Fracchia noted.

After depositing its load of potash, the cooled brine has to be reheated before being pumped back into the mine. That makes the process fairly energy-intensive. While natural gas provides most of the energy, a heat-exchange system transfers some of the heat from the warm brine pumped up from underground to the cooler brine about to be recirculated.

The only one of its kind

The Patience Lake mine wasn't—and isn't—the only solution potash mine, but it is the only conventional mine of that size that has been converted to one. The other solution mine in Saskatchewan is Mosaic's mine at Belle Plaine. "The chemistry is the same, they just use a different method," Fracchia said. "In their case they didn't actually have an active mine, they just had an ore body. They started mining by injecting water and circulating water through the formation. That water . . . started to open up a cavern. The potash that was dissolved was brought to the surface."

The extraction method is also different. "They didn't use outdoor ponds. They used large evaporators that would evaporate off the excess water, and crystallizers to recrystallize the potash, in a plant rather than outdoors. They've since added production ponds as well to supplement their production."

A small mine near Moab, Utah, previously mined conventionally for a few years, had also become a solution mine before the Patience Lake mine converted. Then owned by Texas Gulf, the low-production operation had only one injection well and one recovery well, and used evaporation rather than crystallization ponds (because of the different climate). "It wasn't really comparable," said Fracchia.

"We worked with some consultants," he said, "but in the end we did a lot of (the engineering studies) in-house, because the technology just wasn't there, the experts just weren't there . . . That mining technology really didn't exist anywhere in the world."

Now, of course, it does . . . in Saskatchewan.

"We not only have big ore bodies, we have these high-grade ore bodies, which are stupendous resources."

Uranium

It's not as famous as the ones in California and the Yukon, but Saskatchewan had its own major gold rush, spurred by the 1934 discovery of gold on the north shore of Lake Athabasca, between Lodge Bay and Neiman Bay. The town of Goldfields was one result.

The discovery didn't really (if you'll pardon the expression) pan out. Only one mine grew out of it, and it produced for only three years, from 1939 to 1942, before being shut down due to low-grade ore. When the mine died, so, pretty much, did Goldfields.

But in 1935, while looking for gold, prospectors of the Mineral Locators Syndicate came across something else at the Nicholson property 3 km east of Goldfields: pitchblende.

Pitchblende is primarily uraninite, a mineral consisting of triuranium octoxide (U_3O_8) and trace amounts of radium, thorium, polonium, lead and helium. It's the main ore from which uranium is extracted.

The prospectors recognized pitchblende because they were familiar with the Port Radium radium mine on the Northwest Territories' Great Bear Lake. F.J. Alcock, a geologist with the Geological Survey of Canada (GSC), collected samples of the Lake Athabasca mineral, and GSC mineralogist H.V. Ellsworth positively identified it as pitchblende.

The find attracted little attention at the time. The prospectors were looking for gold. Uranium wasn't worth very much. In fact, since little gold was found on the Nicholson property, it remained closed until 1949.

Uranium becomes valuable

In the intervening years, the value of uranium increased. With the world at war, the United States embarked on a top-secret scientific and military effort, the Manhattan Project, to develop a powerful new weapon: an atomic bomb. In 1942 the Canadian government reopened the Port Radium mine (which had been closed in 1940 and in the interim had flooded and frozen) and its accompanying refinery to mine and process uranium for the Manhattan Project. At the same time as the mine reopened, its owner, Eldorado Gold Mines Ltd., changed its name to Eldorado Mining and Refining Ltd.

Because of uranium's military value, in 1943 the federal government issued a ban on the staking and mining of radioactive minerals by the private sector. In 1944, it expropriated Eldorado, turned it into a Crown corporation, and gave it the sole exploration rights for new radioactive minerals. That same year an intensive effort began to find and develop new sources of uranium. (At the time, the Great Bear Lake deposit, the find near Goldfields and an unconfirmed occurrence near Lake Superior were the only known locations of uranium minerals in the country.)

In September 1944 R. Murphy, exploration manager of Eldorado, and A.W. Joliffe of the Geological Survey of Canada returned to the Nicholson property. After examining it, they recommended prospecting and staking in the area.

During the 1945 season, prospectors armed with Geiger counters (portable radiation detectors) turned up more than 1,000 showings of pitchblende in what became known as the Beaverlodge region. The town of Goldfields revived.

That same summer, on July 16, the United States exploded the world's first atomic bomb and, on August 6, dropped an atomic bomb on Hiroshima, Japan. On August 9, a second bomb was dropped on Nagasaki. The Japanese surrendered, and the Second World War was over.

Left page: A prospector uses a Geiger counter to search for uranium in the rugged terrain around Beaverlodge in 1952. *(Saskatchewan Archives Board R-A2662.)*

Above: Prospectors Ralph Anderson and L.N. McArthur examine uranium ore with W. James Bichan, director of mineral resources, in McArthur's camp at Black Bay, Lake Athabasca, in June 1949. *(Saskatchewan Archives Board R-A9256-2.)*

Right page: Director of Mineral Resources W. James Bichan tries out a prototype drill-hole Geiger counter at Martin Lake in the late 1940s. *(Saskatchewan Archives Board R-A9305-11.)*

The hunt for uranium, however, continued. In September 1946 prospectors discovered pitchblende at Ace Lake and staked the claim for Eldorado. That same year, the federal government set up the Atomic Energy Control Board.

In 1948 the government lifted the ban on private staking, unleashing one of the largest staking rushes in Canadian history. Within five years, more than 200 companies were active in the district, and around 1,000 more occurrences of pitchblende had been discovered.

Of course, merely finding pitchblende doesn't mean you've found the makings of a uranium mine: the deposit has to be big enough, rich enough, and accessible enough to be mined and milled profitably. "That's the definition of an ore body," according to Chuck Edwards, Director of Engineering and Projects for Cameco Corporation. "Otherwise it's not strictly ore, it's an interesting deposit." And, he added, "There are a lot more of them than there are ore bodies."

The settlement of Beaverlodge, Eldorado's base camp, was established in 1949, and Uranium City followed in 1952. The Ace-Fay-Verna mine began production in 1953.

The staking rush continued for a full decade; in fact, 1959 was the peak production year in the Beaverlodge area, with close to 14.1 million kg of yellowcake (milled triuranium octoxide, U_3O_8) being produced. But in 1959 the United States Atomic Energy Commission, which had been purchasing 90 per cent of all Canadian uranium, cancelled all foreign contracts in favour of developing American domestic resources. The contract with Canada expired in 1962. By 1964, Eldorado was the only surviving uranium producer in Saskatchewan—and it only survived because of a federal subsidy and stockpiling program, and by reducing production capacity.

The Ace-Fay-Verna mine, the first to open in the Beaverlodge district, was the last to close, in 1981. Over the years, 16 mines had operated in the area.

New exploration, new finds

The market for uranium began to improve in the late 1960s due to the increased use of uranium as a nuclear fuel for the generation of electric power. That sparked a resurgence in uranium

prospecting. In 1967 a consortium named the Dynamic Group took advantage of a provincial financial assistance program to conduct an airborne survey of the unexplored Athabasca Basin with equipment that could detect sources of radiation. Gulf Oil Company acquired the resulting data, and in October 1968, following up on anomalies in the Wollaston Lake area, drilled the first hole into the Rabbit Lake deposit, from which uranium continues to be produced today. Another land-staking rush followed and lasted until 1972. Then in 1974 a joint-venture group headed by Uranerz Exploration and Mining Ltd. proposed an exploration partnership to the provincial government. The result was the Saskatchewan Mining Development Corporation (SMDC).

In the 1970s, work by geologists with the Saskatchewan Geological Survey—particularly mapping of the Highrock Lake area by G.E. Ray in 1976 and 1977—led to a better understanding of the geology of the Canadian Shield and provided new hints about where to look for uranium. Specifically, Ray's work and follow-up work by other geologists proved that airborne surveys, followed up by ground-based surveys and systematic test drilling with electromagnetic conductors, could locate uranium ore bodies in the deeper part of the Athabasca Basin where there were no surface indicators such as boulders.

By 1978, much of the open ground in the Athabasca Basin was staked. Several significant discoveries followed, but as the price of uranium fell from $40 US per pound of yellowcake in 1979 to less than $10 per pound in 1990, the level of activity fell with it.

In 1982 Eldorado Nuclear Ltd. shut down its Beaverlodge operation, ending all production in the Uranium City area—and effectively ending Uranium City, as well. However, at about the same time Eldorado acquired ownership of the Rabbit Lake mine-mill complex, which had come on stream in October of 1975. In 1988 the federally owned Eldorado and the provincially owned SMDC amalgamated to form Saskatoon-headquartered Cameco Corporation, now the largest producer of uranium in the world.

Recently the price of uranium has been rising, and as a result, claim-staking activity has increased again in the Athabasca Basin. In fact, as of 2005, almost all the open ground, including that in the deep part of the Athabasca Basin, had been taken up.

Three mines today

Currently three mines operate in the province: McArthur River, Rabbit Lake and McClean Lake. A fourth, Cigar Lake, is expected to be on stream by 2007. McArthur River and Rabbit Lake are underground mines, as will be Cigar Lake. McClean Lake, discovered in 1979, is an open pit mine. Owned by COGEMA, Denison Mines and OURD (Canada) Co. Ltd., and operated by COGEMA, it currently produces 2310 tonnes of uranium (about 2700 tonnes of yellowcake) per year.

"Which one you use (open pit or underground), depends on the geology and rock conditions of your ore body," Edwards explained. "Generally if the ore body is close to the surface and there's not a lot of water, then you will open-pit. If you can't open-pit it, or if the ore body is very deep underground, you will go underground."

Rabbit Lake, owned by Cameco, is still in production more than 30 years after it opened. It produces 2077 tonnes of uranium (about

2450 tonnes of yellowcake) per year. It's a conventional underground mine. "There are a variety of underground mining methods, largely determined by the rock characteristics and the geology characteristics you meet underground, and by the ore grade," Edwards said. "As the ore grade rises, at a certain point it's not safe to work with the ore exposed, or in the ore, which is not true, say, for nickel or copper or gold or iron."

Uranium ore is graded by the percentage of U_3O_8 it contains. The Rabbit Lake ore is fairly low-grade, averaging 1.2 per cent U_3O_8. At McArthur River, though, the grade averages an astonishing 25 per cent U_3O_8. "When we developed the mines at Key Lake back in the early '80s," Edwards noted, "the average grade at Key Lake was about 2.5 or 3 per cent, and at that time that was fabulously high-grade. Everyone was astounded, because previous mines were well under 1 per cent. It's unusual in the mining industry to have your grades increasing over time. It usually goes the other way."

High grades require new technology

The high grade ". . . creates problems because if the grade is high enough you have to have an increased level of radiation protection for the workers," Edwards said. "We've developed methods in Saskatchewan for handling and processing these high-grade ores that nobody else uses."

At McArthur River the mining is all done by machine, so the workers are never in or exposed to the radioactive ore. The process is called "raised boring."

"For raised boring, you start above the ore body," Edwards said. "You have openings, tunnels, essentially, above and below the ore body, and you drill a relatively small hole from the upper level through the ore body to the lower level. You put a pipe down that hole, and on the bottom you put a raise bore head, which is essentially a big grinding machine . . . it looks like a 10-foot-diameter pepper mill. You pull that back up, and then turn it as you go, and it basically crumbles the rock, which falls down."

The falling rock is collected by a remote-controlled front-end loader and taken to an underground grinding mill. The grinding mill grinds up the ore and adds water to it to make a slurry, which is then piped to surface for storage, then trucked to the mill for processing into yellowcake.

Cigar Lake, discovered by COGEMA in 1981 and jointly owned by Cameco, COGEMA, Idemitsu Uranium Exploration Canada Ltd. and TEPCO Resources Inc., also boasts impressively high-grade ore, averaging 21 per cent U_3O_8. It's expected to go into operation in 2007 and produce up to 6924 tonnes of uranium (8182 tonnes of yellowcake) per year.

Another entirely new method of mining has been developed for Cigar Lake. "Jet boring" uses a high-pressure water jet operating beneath the ore. The ore breaks up in the stream of water, falls, is immediately slurried, and then is pumped away, fully contained, to a storage area to await grinding in the underground mill.

Although open-pit mining is the cheapest form of mining and conventional mining is the next cheapest, "the benefit for us is, although the automated mining methods are expensive on a per-tonne basis, because the grades are so high they're less expensive per pound of uranium," Edwards said.

Mining methods at McArthur River and Cigar Lake were developed in Saskatchewan and are unique to the Saskatchewan uranium mining industry, Edwards said. "Other people use raised borers, but they use them to make holes in rock; as far as I know, no one else uses them for mining.

"There is only one jet boring mining machine in the world and we own it. We are just going out now for the engineering and construction of the second."

Making yellowcake

Once in the mill, the fine-ground uranium ore is brought into contact with sulphuric acid and an oxidant. This dissolves the uranium. "What you would like to do is dissolve all the uranium and nothing else," Edwards said. "It doesn't happen that way."

The resulting slurry consists of a liquid that contains the uranium, and solid waste known as tailings. The slurry passes through a solid-liquid separation stage. The tailings go into a tailing management facility for permanent storage, while the remaining solution goes through a purification stage in which an "organic extractant"

Left page: The high-grade ore at McArthur River is too radioactive for workers to mine it directly. Here raise bore operator Al Tuttle adds a length to the drill as a pilot hole is drilled from the 530-m level through the ore zone down to a tunnel on the 640-m level. *(Photo courtesy Cameco.)*

Left: Rabbit Lake, the longest-producing uranium mine in Saskatchewan. *(Photo courtesy Cameco.)*

attracts the uranium and leaves the impurities behind in the waste water, called raffinate. The raffinate is cleaned, then discharged.

The uranium is then removed from the organic extractant via a process called stripping, which brings the uranium back into a water solution. "What you end up with then, if you've done it right, is a solution of virtually pure uranium at a higher concentration," said Edwards.

The next stage is precipitation: the uranium is precipitated out of the solution by the addition of either ammonia (as at Key Lake) or hydrogen peroxide (as at Rabbit Lake). The resulting yellowcake is dried, then shipped to refineries in Ontario for the next stage of processing into nuclear fuel for reactors.

Left: Key Lake is the world's largest high-grade uranium milling operation. *(Photo courtesy Cameco.)*

Right page: Cigar Lake is the world's largest undeveloped uranium deposit. *(Photo courtesy Cameco.)*

Innovations in the mills

From 1983 to 2000 the mine at Key Lake was the largest in the world. Although the mine closed in 2000, the mill has remained in operation processing the ore from the McArthur River mine, located just 80 km to the northeast. Cameco's other mill is at Rabbit Lake.

The Key Lake and Rabbit Lake mills both boast unique, environmentally friendly innovations. When ammonia is used, one by-product is ammonium sulphate. "Prior to the Key Lake mill being built . . . the sulphate was simply put out into the environment," Edwards said. "The Key Lake mill was developed with a method of trapping this ammonium sulphate product, and we crystallize it." The crystallized ammonium sulphate is sold to companies for use in fertilizer. "That was a major innovation and it cut down the environmental impact of a uranium mill a huge amount."

The Rabbit Lake mill uses concentrated sulphuric acid instead of ammonia for stripping, and hydrogen peroxide rather than ammonia for precipitating. "We don't have an ammonium sulphate trapping circuit at Rabbit Lake because there is no ammonia in the process. That's quite an elegant process." Hydrogen peroxide in particular, noted Edwards, "is a very nice reagent because its by-product is water."

The pre-eminent producer

Canada is the world's largest producer of uranium, accounting for 30 per cent of world production —and all of Canada's uranium comes from Saskatchewan. In 2004, that uranium was worth an estimated $800 million. Uranium is the second most important mineral resource for the Saskatchewan economy, behind potash.

Saskatchewan holds a pre-eminent place in world uranium production because "we not only have big ore bodies, we have these high-grade ore bodies which are stupendous resources," Edwards said. "We are a huge producer. Huge."

How huge?

One year's production of uranium from Saskatchewan contains more energy than all of Canada's annual oil and natural gas production.

Now that's huge.

"There is every reason to believe that yes, there will be a diamond mine in the Fort-à-la-Corne forest."

Diamonds Are a Province's Best Friend

Earlier in this chapter it was pointed out that Saskatchewan was a very different place 380 million years ago. Well, it was also a very different place 100 million years ago when, not far from where Prince Albert stands today, volcanoes were erupting.

The molten rock that forced its way to the surface was kimberlite, which contains a lot of "volatiles"—gases. Magma with a lot of gas in it causes a volcano to erupt explosively, like a warm bottle of champagne popping its cork.

However, it's not the gases in kimberlite that endear it to geologists: it's the diamonds. Formed from carbon by the immense pressures 150 km or more below the earth's surface, the diamonds come along for the ride when kimberlite wells up as it did in Saskatchewan a thousand millennia ago. And thanks to that geological event, it seems likely that the province will have one, and perhaps two, producing diamond mines within the next decade.

First diamonds found in 1948

The first diamonds discovered in Saskatchewan were not found through multi-million-dollar efforts like the ones now under way: they were five stones that a prospector reported he had found in the Prince Albert area in 1948 (although his find was never confirmed).

In 1961, another prospector said he had recovered two diamonds, each about 0.6 cm in diameter, in gravel 9.6 km west of Prince Albert. His claim, too, was never substantiated, but Dr. Mousseau Tremblay, employed by a subsidiary of the South African diamond giant De Beers, toured the area and found kimberlite-related rocks.

Left page: George H. Read, Vice President of Exploration for Shore Gold Inc., shows off one of the diamonds recovered from the Star kimberlite. *(Photo courtesy Shore Gold Inc.)*

Left: Inside one of the wetter tunnels in the extensive network Shore Gold has dug 235 m underground in the Star kimberlite. *(Photo courtesy Shore Gold Inc.)*

Those possible finds were interesting and suggestive, but the fledgling Saskatchewan diamond industry was really born in 1988 when De Beers dug a hole at Sturgeon Lake, about 30 km northwest of Prince Albert, and found kimberlite. Among those on hand for that test hole was George H. Read, who is now Vice President of Exploration for Shore Gold Inc., the company which today leads the way in the pursuit of Saskatchewan diamonds.

In 1989, Klaus Lehnert-Thiel and Jiri Strnad and others, working for Uranerz Exploration and Mining Ltd., examined a series of Geological Survey of Canada regional aeromagnetic maps and identified several circular geomagnetic anomalies in the Fort-à-la-Corne Provincial Forest, 80 km east of Prince Albert. They thought those anomalies could be kimberlite, Read said, "because they were magnetic anomalies in a sedimentary basin, and it's very hard to account for them other than by some form of volcanic rocks such as kimberlite."

He added, "I think they even discussed the subject with De Beers. De Beers at that time didn't think they were kimberlites because they were too big, which was a very reasonable conclusion at that time."

A follow-up, more detailed aeromagnetic survey identified approximately 70 anomalies, and in August of 1989 Uranerz did some test drilling and found kimberlite. Claims were staked, with Cameco and Uranerz 50/50 joint venture partners in the prospecting.

De Beers joined the joint venture in the early '90s. Cameco eventually acquired Uranerz, and Kensington Resources Ltd. became part of the joint venture, with 42.245 per cent each held by Kensington and De Beers and the balance by Cameco. Then, in October of 2005, Shore Gold acquired Kensington, giving it 42.245 per cent of the joint venture.

From 1990 to 2000, the Fort-à-la-Corne Joint Venture recovered 1,249 macrodiamonds (greater than 0.8 mm in diameter) with the

largest being 1.535 carats. (One carat is equal to 0.2 g.) In 2000, De Beers mineral resource evaluation department reviewed the data and chose two kimberlites (of some 69 that had been identified) for bulk sampling with large-diameter drills. By the end of 2005, approximately 366 carats of diamonds have been recovered by the Fort-à-la-Corne Joint Venture.

But that's not the only diamond project under way in the region, and in fact the other is probably further advanced.

The Star Diamond project

Adjacent to the southeast end of the array of kimberlites that make up the Fort-à-la-Corne Joint Venture lies Shore Gold's wholly owned Star Diamond project. That parcel wasn't included in the original claim stake for the Fort-à-la-Corne Joint Venture because it was held as Crown reserve land. When the land was released in 1995, Shore claimed it.

"I think initially it was suspected they had a number of kimberlites in the ground, and then once some 40 holes were drilled as part of the exploration program, they realized they had one very big kimberlite called the Star kimberlite," Read said.

All the kimberlites in the area are buried under about 100 m of overburden, deposited by glaciers. "It makes their exploration slightly more complicated," Read said. He explained that in order to truly evaluate a kimberlite, it's necessary to recover a large parcel of diamonds.

Unlike gold, which is worth the same per unit of weight no matter where it's found, diamonds are unique to the kimberlite from which they are recovered. "You could have a diamond population that has a high value in one kimberlite and a low value in another kimberlite," said Read. That means that to adequately evaluate a potential mine site you need to get a large parcel of diamonds from it, from 2,000 to 3,000 carats. Shore decided the best way to obtain that many diamonds was to sink a shaft.

Finding the diamonds in the kimberlite haystack

George Read joined Shore in October of 2003, just when the shaft was being sunk. Much of the material overlying the kimberlite is "very unconsolidated and therefore unstable, and difficult for miners to sink a shaft through," he said, "and so the ground was frozen with a freeze wall . . . a number of drill holes were drilled to the depth of 130 metres below ground, and

Left page: A worker drills into the wall of a tunnel 235 m underground in the Star kimberlite. *(Photo courtesy Shore Gold Inc.)*

Left: Examining a kimberlite sample deep underground. *(Photo courtesy Shore Gold Inc.)*

Above: Deep underground in the Star kimberlite. *(Photo courtesy Shore Gold Inc.)*

Right page: One of the high-tech rigs made by Bauer Engineering currently drilling a network of 72 1.2-m-diameter holes to a depth of 360 m in the Star kimberlite to determine the grade distribution across the formation. *(Photo courtesy Shore Gold Inc.)*

connected to a freeze plant, very similar to the freeze plant that would be used on a commercial ice hockey rink. The ground was frozen down to 130 metres, and then blasted through." The resulting shaft was then walled with concrete.

"We entered kimberlite in my second week of employment with Shore, 107 metres below surface, and then the shaft was sunk down to eventually 250 metres below surface," Read recalled. "On the 235-metre level we established a very large network of tunnels, lateral drifts, as we call them. They're about 2.4 by 2.4 metres in size. From that we extracted kimberlite in 250- to 350-tonne batches."

While all that was going on, a large on-site diamond processing plant had been established. "A diamond processing plant is very environmentally friendly," Read said. "It is a process of crushing, scrubbing and screening, and recovery of diamonds is based on the physical characteristics of diamond." Specifically, diamonds are dense, fluoresce in X-rays, and stick to grease.

Once the kimberlite is crushed, scrubbed and screened, a fraction of it, between 16 and 22 mm in diameter, is fed into a "dense media separator," a cyclone filled with a spinning ferrosilicon slurry. The dense minerals such as diamonds—anything denser than 3 g per cu. cm—sink. The light minerals float. The resulting heavy mineral concentrate, a very small fraction (something like 0.5 per cent) of the kimberlite that was poured in at the head end of the plant, is next fed through the "X-ray flow sort," where it is bombarded with X-rays. When the telltale X-ray fluorescence of a diamond is detected, a trap door opens and the diamond falls into a secure container.

The remaining material, still wet, is fed over an inclined, vibrating table covered with a very sticky grease. "Diamonds are said to be hydrophobic or oleophilic—in other words, they are not wettable," Read explained. "The concentrate that is wet does not stick on this inclined, vibrating grease table, whereas the diamonds do stick in the grease."
The material collected from the grease table and the X-ray sort is then placed in sealed packages, given to secure carriers, and sent off to Ontario to be individually picked, weighed and described.

Good valuation means more financing

"We eventually took out some 27,800 tonnes," said Read, "and from that we recovered 4,047 carats (80.94 g) of diamonds. As soon as we had 3,050 carats we proceeded with an evaluation exercise, which meant exporting the diamonds to Antwerp."

In Antwerp, Belgium, four expert companies examined the diamonds and determined an average value. The results were good: whereas the average value of diamonds sold worldwide is $70 US per carat, the diamonds from the Star kimberlite were valued at $135 per carat.

"We are also very encouraged by the size of the diamonds that we see at Star," Read said. "The largest stone that we recovered in the bulk sorting operation was 19.67 carats, and that was probably limited by the top screen of the plant."

Not only that, about 70 per cent of the diamonds are gem quality, he said. Many of them are very good white stones, and the octahedral shapes favoured by gem cutters are prevalent. "A positive feature is that there appeared to be some vivid or fancy yellow diamonds as well. We have recovered two to date," Read said.

The positive valuation of the Star kimberlite diamonds brought in investment from Newmont Mining Corporation of Canada, Ltd., the world's largest gold producer. "I think Newmont was attracted by the size of the Star kimberlite and the quality of the diamonds," Read said. "They are also quite familiar with the incredible economics of a world-class diamond mine, should Star get to that position. The cash flow from a large diamond mine is exceptionally good. The Diavik diamond mine, established some 300 kilometres north of Yellowknife, had capital costs in excess of $1 billion, and a repayment period of 32 months on capital, so an extremely good cash flow."

In March 2005 Shore announced a $116-million financing, of which Newmont took $51 million, which gave them a 9.9 per cent holding in Shore Gold. With that financing, Shore was able to proceed with a

Saskatchewan kimberlites: unique in the world

As George Read noted, when De Beers—the world experts in diamond mining—first looked at the findings of Klaus Lehnert-Thiel and Jiri Strnad, they thought that the geophysical anomalies were too big to be kimberlites.

But when the Fort-à-la-Corne kimberlites erupted, they were on the eastern edge of the Western Interior Seaway—and it was growing. The first phases of kimberlite eruption were probably on the shore, but later eruptions probably occurred in shallow water.

Normally a kimberlite is a pipe-like body with a fairly limited surface area, because the material that exploded and poured out onto the surface during the volcanic eruption has long since eroded away. "The minerals that are formed at depth in the mantle at high temperatures and pressures fall apart at the Earth's surface, and the diamonds that might be in that rock are then released into the surrounding streams and lakes," Read explained.

But because of the expanding sea, the Fort-à-la-Corne kimberlites were protected, first by water, then by sediment. As a result, the whole of the "volcanic edifice," rather like the cap on a mushroom, with the pipe being the mushroom stem, was preserved.

"The cap of the mushroom was what caused this big geophysical anomaly, as opposed to the very small pipe underneath," said Read. That hasn't happened anywhere else in the world where kimberlite is found, he said.

"It is unique."

pre-feasibility study on the Star kimberlite. As part of that study an additional 15 000 tonnes of kimberlite has been recovered. As of December 2005 more than 8000 tonnes of that had gone through the processing plant, producing a further 1,000 carats (20 g) of diamonds. The goal is to increase the carat parcel from 4,000 to 6,000 carats.

Another component of the pre-feasibility study is a detailed core drilling program. Three-inch (7.62 cm) diameter cores are being drilled on a 100-m grid pattern across the central part of the kimberlite, and a 200-m grid on its periphery. "The aim of the core drilling is to define the shape, size and internal structure of the Star kimberlite," Read said.

Following on the core drilling, two high-tech rigs made by Bauer Engineering in Germany are drilling holes 1.2 m in diameter to a depth of 360 m. Eventually there will be 72 of these holes. "They will be drilled adjacent to known core holes so we will know the geology we are entering," Read explained. "The aim of that is to determine the grade distribution across the Star kimberlite. Once we have the internal structure and the grade distribution, we can construct a 3-D model."

The pre-feasibility study should be complete by the end of 2007, after which some additional engineering work will be required to raise it to a "bankable" feasibility study, which can be used to borrow additional funds needed to build an operating diamond mine. "Hopefully that construction will happen by 2009 and 2010 with potential production in 2011," Read said.

"The outlook is very good," he summed up. "Obviously we need to prove these projects as diamond mines. We have to go through the rigorous process of permitting and environmental impact assessment. We have to do it right and do it right the first time.

"Certainly the results to date are very encouraging. There is every reason to proceed and there is every reason to believe that yes, there will be a diamond mine in the Fort-à-la-Corne forest."

The rediscovery of Avonlea-Truax bentonite

Sodic bentonite is a type of clay that swells to many times its dry volume when wet. This (along with certain other properties) makes it extremely useful in a number of industrial capacities: for example, to clarify beer and wine, to bind animal feed, as a component of pond liners, to thicken drilling mud, as an absorbent in materials like kitty litter, to name just a few.

Terence Field was employed by Saskatchewan Economic Development and its precursor, Saskatchewan Industry and Commerce, from 1974 to 1996. "For most of that period one of my sectors was the clay and clay products industry," he wrote to the author. "As a result I became aware of the story behind the rediscovery of the Avonlea-Truax bentonite."

In the early 1970s John Hudson travelled to the rugged terrain in the Prairie Farm Rehabilitation Administration (PFRA) community pasture in the Avonlea-Truax area on one of his periodic treks around the province in search of rare plants. Botany may have been Hudson's passion, but his scientific specialty was clays and ceramics. A graduate of the University of Saskatchewan, Hudson had been employed by the Saskatchewan Research Council since the late 1950s.

"He quickly lost interest in botany . . . when he noted three bentonite seams, each about 18 inches (45.7 cm) thick," Field wrote. "Due to his own clay research work, and having visited numerous clay outcrops around the province, he was aware of other bentonite occurrences, such as at St. Victor and Pickthall, and instantly realized that the outcrops he was now seeing were far superior to the others. He had some sample bags with him and took samples back to his office in Saskatoon for testing."

Although Hudson was apparently the first to realize the importance of the bentonite beds, they had actually been spotted many years earlier. Back in 1937 federal government geologists identified a dome-like structure in the Avonlea-Truax area that they thought might be a trap for petroleum. They used the bentonite beds as a marker formation in their mapping of the area. Field wrote that Hudson believed Saskatchewan Natural Resources was aware of the federal geologists' discovery of bentonite in that area, but hadn't followed up on it.

"Shortly after Hudson sampled the Avonlea-Truax bentonite, he received an unexpected visit from Vern Hogg Jr., principal of a Calgary exploration firm, Flin Flon Mines," Field wrote. "As a result of his conversation with Hudson, Hogg initiated exploration work in the Avonlea-Truax area. Flin Flon Mines eventually took out a lease in the area, but others had become aware of the activity, and also leased acreage in the area. The other party subsequently sold their leases to Inland Cement, and some years later Flin Flon Mines sold its leases to Saskatchewan Minerals."

When a local recession resulted in the Inland Cement plant being taken out of cement production for several months in 1972, the plant dried and ground several thousand tonnes of bentonite, marketing it to iron ore palletizing plants in northwestern Ontario, Field wrote. When cement market conditions improved and the plant reverted to cement production, the bentonite leases were made available through a royalty arrangement to some Inland employees, who organized a bentonite project. Shares were sold locally, and the newly organized company, called Avonlea Minerals Industries Ltd., established a quarry near Truax and, in 1978, an 80 000-tonne-per-year processing plant at Wilcox.

Avonlea became a subsidiary of Ekaton Industries (another venture of Vern Hogg Jr.) in the late 1980s, but both Ekaton and Avonlea became insolvent in 1990. Canadian Clay Products Inc. resumed operation of the quarry and plant in 1991.

CHAPTER 7 SOURCES

Books

Kupsch, W.O. and Hanson, S.D., ed. *Gold and Other Stories as told to Barry Richard*. Regina, 1986. Saskatchewan Mining Association.

Storer, Dr. John. *Geological History of Saskatchewan*. Regina, 1989. Saskatchewan Museum of Natural History.

Websites

Cameco Corp. *Uranium 101*. (March 18, 2006.) < http://www.cameco.com/uranium_101/>

Cogema Resources. (March 18, 2006.) <http://www.cogema.ca/>

De Beers Group – Canada - Fort à la Corne. (March 18, 2006.) <http://www.debeersgroup.com/debeersweb/About+De+Beers/De+Beers+World+Wide/Canada/Fort+%C3%A0+la+Corne.htm>

PotashCorp. (March 18, 2006.) <http://www.potashcorp.com>

Saskatchewan Centre for Soils Research, College of Agriculture, University of Saskatchewan. *Saskatchewan Interactive: Mining*. (March 18, 2006.) <http://interactive.usask.ca/ski/mining/index.html>

Saskatoon Airport Authority – Facilities & Services – Entertainment. (March 18, 2006.) <http://www.yxe.ca/facilities/entertainment.php>

Shore Gold Inc. (March 18, 2006.) <http://www.shoregold.com>

Other sources

Field, Terence. Letter to the author. July 6, 2005.

Harper, C.T., ed. *Geology, and Mineral and Petroleum Resources of Saskatchewan*. Regina, 2003. Saskatchewan Geological Survey, Saskatchewan Industry and Resources. Miscellaneous Report 2003-7.

Vigrass, Laurence. "Elk Point Sea." *The Encylopedia of Saskatchewan*. Regina, 2005. Canadian Plains Research Center.

Personal interviews conducted by the author

Edwards, Chuck. Director of Engineering and Projects, Cameco Corp. November 18, 2005.

Fracchia, Mark. General Manager, Lanigan mine, PotashCorp of Saskatchewan. November 25, 2005.

Read, George H. Vice President of Exploration, Shore Gold Inc. December 4, 2005.

STOCK PHOTO

CHAPTER 8

Urban Engineering

“It was indeed a hand-rubbed piece of community furniture.”

Above (left): In July of 1932, a wooden support arch takes shape as construction proceeds on the Broadway Bridge. *(Photograph A185 by Leonard A.J. Hillyard courtesy Saskatoon Public Library - Local History Room.)*

Above (right): The men responsible for overseeing the design and construction of the Broadway Bridge pose in front of their creation, also known as the Engineers' Bridge. In the centre of the front row is C. J. Mackenzie, dean of engineering at the University of Saskatchewan, who spearheaded the bridge's construction. *(Photograph A2615 by Leonard A.J. Hillyard courtesy Saskatoon Public Library - Local History Room.)*

Right page: Workers install steel reinforcing rods in the Broadway Bridge deck in September 1932. Because it was a make-work project intended to alleviate unemployment during the Great Depression, the Broadway Bridge was deliberately built in the most labour-intensive way possible. *(Photograph A199 by Leonard A.J. Hillyard courtesy Saskatoon Public Library - Local History Room.)*

The Engineers' Bridge (and Others) of Saskatoon

In a word association game, the word most people would associate with the phrase “engineering project” would probably be “bridge.” Maybe it's because bridges wear their engineering proudly, flaunting the arches or cables or piers that support them.

Saskatchewan's huge road network boasts thousands of bridges. But only one community has earned the nickname “City of Bridges,” and only one bridge in that city has earned the sobriquet “The Engineers' Bridge.”

The city is Saskatoon, and the bridge is the Broadway Bridge.

The Dean's idea

Dr. C.J. Mackenzie, the first Dean of Engineering at the University of Saskatchewan, and the first president of the Association of Professional Engineers of Saskatchewan, had a strong interest in town planning.

In 1928 Mackenzie became chairman of the City Planning Board. A consultant to the board, A.E.K. Bunnell, suggested in a study that a downtown commercial area at risk of becoming a slum could be saved by linking it to Broadway Avenue via a bridge. Mackenzie liked the idea and began promoting it. Mayor George W. Norman told him that if he really wanted the project adopted, he needed to speak "from the seat of power"—in other words, he needed to get himself elected to city council. (Macdonald, p. 60.)

Mackenzie did exactly that. He served just one year (1928-29), but during that year he won acceptance for the new bridge and saw the necessary bylaw passed. Despite that, the bridge seemed years from construction.

Then came the crash of 1929 and the Great Depression. People struggled to survive. To alleviate the devastating level of unemployment, the federal government approved a number of "make-work" projects—including Mackenzie's Broadway Bridge.

Named head of the project, Mackenzie took a leave of absence from the College of Engineering to build the bridge. His design and engineering staff comprised recent university graduates: Kent Phillips, Bev Evans (the chief draftsman), Bill Cumming (who had worked with

STOCK PHOTO

Now, that's using your (daughter's) head

According to *Bridge Tales*, a video about the bridges of Saskatoon prepared by the Meewasin Valley Authority, one element of the Broadway Bridge's design came about in an unusual way.

One day while Mrs. Mackenzie and her daughter were crossing the University Bridge, the little girl kept sticking her head through the railings to see the river water rushing below. When the two got home, Mrs. Mackenzie told her husband in no uncertain terms that the railings on the new bridge must be made safer.

Mackenzie promptly took his daughter to the kitchen, had her put her head in the door, and then closed the kitchen door just enough to get a measurement of the width of her head.

He then designed the railings on the Broadway Bridge to be too small for a head that size to poke through. *(Bridge Tales.)*

Above (left): Wagons cross Saskatoon's Traffic Bridge (later known as the Victoria Bridge) in 1912 under the watchful eye of a police officer. *(Photograph LH5080 by Leonard A.J. Hillyard courtesy Saskatoon Public Library - Local History Room.)*

Above (right): Saskatoon residents examine the wreckage after an outbound train derailed on March 12, 1912, destroying an entire span of Saskatoon's first railroad bridge. Although a sleeping car fell 15 m to the ice, injuring a dozen people, no one was killed. *(Saskatchewan Archives Board R-A2015.)*

Right page: The wreck of the steamer *City of Medicine Hat* lies against one of the Traffic Bridge piers on June 7, 1908. *(Photograph LH308 courtesy Saskatoon Public Library - Local History Room.)*

the Dominion Bridge Company) and Bob Donald. (Cumming and Donald supervised the steel reinforcement, which was shaped on site under their direction.)

Gordon Parkinson, a recent recipient of a Master of Science degree in the United States, was in charge of the deck work. Rounding out the engineering team were Clarence Forsberg, a college instructor, and Clyde Jackson.

Several city engineers also had a hand in the bridge. Wally Foss represented Saskatoon as resident engineer, Merv Carson did the surveying, Howard Douglas worked as an engineer with the contractor (R.J. Arrand of Saskatoon), and Bill Rieson was a rodman. (Macdonald, pp. 61-62.)

Work started on December 12, 1931. To maximize employment, no machinery could be used for work that could be done by hand, all labour had to be requisitioned from the local relief office, and most labourers were allowed to work for only one week. In all, 1,593 men were employed. A childless man could not earn more than $25 a month, while the maximum for a man with a large family was $37.50.

The piers had to be completed before the spring floods, which meant 18-hour working days for both the labourers and the four design engineers. Working in the fourth-year drafting laboratory at the university, they had to have working plans ready each morning. (Macdonald, pp. 60-61.)

The final operation was the scouring of the structure, budgeted at 25 cents per square foot. Workers used water and carborundum to scrub off all the rough spots and any imprints left in the concrete by the wooden forming. As author R.H. Macdonald wrote in *Thorough*, his history of the University of Saskatchewan's College of Engineering, "It was indeed a hand-rubbed piece of community furniture."

The bridge was officially opened on Remembrance Day, 1932. It was 396 m long, including approaches, and cost less than $1 million to build. Promptly dubbed "The Engineers' Bridge" by University of Saskatchewan students, it was considered "one of the finest structures of its kind in Canada." (Macdonald, p 61.)

Saskatoon's other bridges

Saskatoon is called the "City of Bridges" because it's home to seven bridges within the city limits.

Before the bridges, a cable ferry took passengers and goods across the river. That was less than satisfactory: the saying at the time was that the ferry didn't run when the water was high, it didn't run when the water was low, it didn't run when the wind was blowing and it didn't run when the wind wasn't blowing. And if none of those conditions existed, the ferryman was busy doing something else. The easiest time to get across the river was in winter, when it was frozen. *(Bridge Tales.)*

Saskatoon's first bridge was a rail bridge built in the spring of 1890. Just a year later, ice from the spring breakup piled up against the bridge, and on April 10 it collapsed.

The bridge washed out several more times over the next few years. A washout in 1905, just at the peak of the influx of new settlers into the province, may have slowed Saskatoon's growth. "A lot of people passed by, aware of Saskatoon only because they had heard it was difficult to get to." *(Bridge Tales)*

On March 12, 1912, an outbound train derailed as it crossed the rail bridge, taking out an entire span. One sleeping car crashed 15 m to the ice, injuring 12 people.

The Capsizing of the *City of Medicine Hat*

On June 7, 1908, the sternwheeler *City of Medicine Hat* crashed into a pier of Traffic Bridge. The current swung the boat around hard against the pier's sharp concrete prow, splitting open the hull. Tonnes of white flour poured into the river, but fortunately the passengers all escaped by climbing onto the bridge. The only traffic on the bridge at the time, a herd of cattle, fled the scene in a stampede. *(Bridge Tales.)*

"The greatest marine disaster in the history of Saskatoon" spelled an end to large-scale steamer traffic on the South Saskatchewan River.

Above: Workmen putting the reinforcing rods in place on the form work of the arches for the University Bridge, July 6, 1914. Plagued by construction and financial problems, the bridge didn't open until 1916. *(Saskatchewan Archives Board R-A341-15.)*

Right page: Boaters on Wascana Lake watch construction of the Saskatchewan Legislative Building, ca. 1909. *(Saskatchewan Archives Board R-B9442.)*

Time for a traffic bridge

People could cross the rail bridge if no trains were coming, but that obviously wasn't the best solution. Early in the new century a push began for a "traffic bridge," designed for city traffic only. By 1906 the funds were in place.

The new bridge was originally called the Camelback bridge because of its "humps"—arches in the steel support beams above the deck, which rests on concrete piers. The bridge, draped in red, white and blue bunting for the occasion, was officially opened on October 10, 1907.

Called "Traffic Bridge," "Iron Bridge," "19th Street Bridge," and finally "Victoria Bridge," Saskatoon's first non-railway bridge closed late in 2005 after engineers discovered severe corrosion. (CBC News.)

Two more rail bridges were built in 1907 and 1908. First came the Grand Trunk Pacific Bridge south of town, near the present-day Queen Elizabeth Power Station. The piers went in first, then, over the winter, a crane built up the steel, travelling back and forth across the river on the ice.

Once that bridge was completed, the crew moved downstream to work on a new Canadian Pacific Railway (CPR) Bridge. Its construction featured a thick cable across the river along which a giant bucket moved, carrying up to 10 tons of concrete at a time. Although the bridge was reinforced so a traffic lane could be added to it, that never materialized.

As the city boomed from 1910 to 1913, work began on a new traffic bridge at 25th Street. Seven of the nine piers were in place by 1913, but the boom ended that year, and the next year war broke out. Construction almost halted, amid rumours the contractor had been cutting corners to maximize his profit. An investigation delayed construction further. The University Bridge finally opened in 1916.

More than three decades passed after construction of the Broadway Bridge before Saskatoon got another new bridge. On October 5, 1964, the last train crossed Saskatoon's first railway bridge, which was then pulled down. On October 28, 1966, the Idylwyld Bridge opened, replacing the downtown rail link with a downtown freeway link.

Saskatoon's most recent bridge, the Circle Drive Bridge, was officially opened on July 1, 1983.

New bridges will likely be needed as the city grows—which means that in the future, as in the past, the bridges of Saskatoon will continue to excite interest—and employ engineers.

"It was really a heavy mining operation going on right in the heart of the city."

The Big Dig

Regina's downtown lake began life as a shallow, 65-ha reservoir with an average depth of 1.5 m, created for the most prosaic of reasons: to water livestock.

The federal government and Canadian Pacific Railway formed the reservoir by damming Wascana Creek just west of where the Albert Street Bridge stands today. The notion of turning the erstwhile stock pond into a landscaped centrepiece coincided with the deci sion to locate the Saskatchewan Legislative Building on the south side of the lake. A Montreal landscape architect, Frederick Todd, drew up plans for the surrounding grounds. He recommended the lakeside be beautified with vegetation. By late spring of 1913, 11,000 trees and shrubs had been planted.

Above: Wascana Lake as it appeared in the years before the Big Dig. *(Photo courtesy Wascana Centre Authority.)*

Right page: This map details the work undertaken as part of the Wascana Lake Urban Rehabilitation Project. *(Map courtesy Wascana Centre Authority.)*

But when drought struck in 1930, the shallow lake practically ceased to be a lake at all. The water evaporated and Wascana Creek slowed to an even more sluggish trickle than usual.

Many hands, but hard work

The provincial minister of public works, Jimmy Bryant, suggested deepening the lake as a make-work project, much like Saskatoon's Broadway Bridge. Digging began on September 8, 1931. To maximize employment, use of heavy equipment was prohibited: the workers shovelled dirt by hand into horse-drawn dump wagons during two daily shifts, from 6 a.m. to 12:30 p.m. and from 12:30 p.m. to 7 p.m.

"I met this fellow who was working on the job," Clarence Taylor, who remembered the 1931 project from his childhood, told author Bob Hughes for his book *The Big Dig: The Miracle of Wascana Centre*. "They were so happy just to get work. He said it only took them two months to get it done and get those islands built. . . . They'd work from dawn to dusk and only stop for lunch and to feed the horses." (Hughes, p. 29.)

The project employed 2,107 men, who moved more than 91 000 cu. m of earth, deepening Wascana Lake by two-thirds of a metre. Labourers earned 40 cents an hour; a man and team earned 70 cents. The City of Regina paid $25,712.12 of the total cost of $102,848.51, and the provincial and federal governments paid the rest.

The workers also built two islands, eventually named Willow and Spruce. (Area farmers were paid $3 per yard for rocks for the islands' sloping sides.) "The formation of the islands was Jimmy Bryant's idea," Taylor said. "At the time, he was heavily criticized for it. They thought he'd lost his mind. Oh, he caught heck for that . . . They called it Bryant's folly." (Hughes, p. 28.)

Oddly enough, the project doesn't seem to have attracted much attention. As Taylor said, "I don't remember folks stopping to watch it. What was there to see? A bunch of men shovelling dirt? You could see that in your backyard." (Hughes, p. 29.)

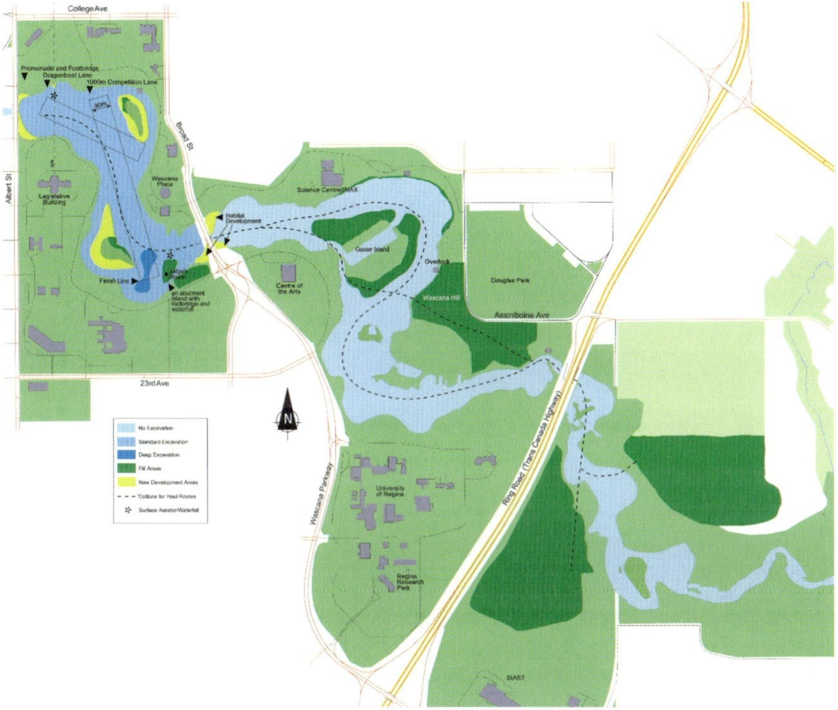

When the time came to re-deepen the lake, early in the 21st century, that would change.

"An ideal marsh habitat"

From an engineering perspective, the "Big Dig" began with Blair Paterson, former executive director of Wascana Centre Authority. In the late 1990s, Paterson convinced the authority to do a feasibility study. "Blair had long had the vision of deepening the lake," recalled Wayne Clifton, President of the engineering firm Clifton Associates, which headed up the study. "We put together a multidisciplinary team, principally geotechnical, but (including) biologists and civil engineers. We studied first the biophysical components of the lake and why it was morphologically changing."

The lake, they found, accumulated sediment (decaying plants and soil from surrounding fields) at a rate of 6 to 8 mm a year. They also confirmed what everyone joked about: the water was extremely shallow. "It was an ideal marsh habitat, but Regina didn't want a marsh, they wanted a lake," Clifton said.

The team concluded that within a couple of decades Wascana Lake would be so choked with vegetation that it could no longer be maintained as a lake. A biologist advised that in order to remain weed-free, Wascana Lake needed to be at least 5 m deep, so sunlight couldn't reach the bottom.

The team settled on a 5.5- to 6-m depth as ideal. Sediment would have to accumulate for about a century before weed growth would become significantly re-established, Clifton said, "making it an aesthetically pleasing water body that could be used for recreation, principally rowing, boating, and so on."

"The stars aligned"

Wascana Centre Authority pushed the idea of deepening the lake, with the support of Doug Archer, then mayor of Regina, but it wasn't until 2002-03, said Clifton, that "the stars aligned." In 2005, Saskatchewan would be celebrating its centennial, and Regina would host the Canada Summer Games,

Above and right page: On March 22, 2004, as excavation ends, the equipment and workers who had taken part in the project gather on the dry lakebed for a ceremony and photographs. *(Photo courtesy Wascana Centre Authority.)*

for which Wascana Lake would be the focus, both for boating events and for general celebrations.

On October 3, 2003, Regina MP Ralph Goodale, federal minister of public works, announced that Ottawa would provide $9 million toward a Wascana Lake rehabilitation project. With federal money assured, the provincial government announced it would provide $5 million and the city said it would provide $4 million.

Well before the official announcement, Saskatchewan Property Management Corporation had been given project management duties. It called for design proposals in June of 2003, and Clifton Associates won the commission to design the project. "We put together a multidisciplinary team," Wayne Clifton said. "It included several geotechnical, structural, hydrology and water-quality engineers, aquatic biologists and landscape designers—quite a broad range of disciplines. Any one team meeting had about 15 or 16 people there, plus owner representatives and their communications people."

Design work started the last week in June. The goal was to complete

the project by the summer of 2004 so the venues could be tested for 2005's Summer Games. It was a very short time line—and it was about to get shorter. "When we started, the owners believed that it was a fully licensed and permitted project and ready to go to construction," Clifton said. "Very soon we found it was not, and that we had to start the environmental assessment and permitting game all over again. So it was a very major rush."

Early on, the design team identified other opportunities for improving Wascana Centre "rather than simply digging the dirt out of the lake and dumping it somewhere." One goal was to improve the strip of land on the lake side of Albert Street Bridge, which, Clifton said, "was really a mess. It was full of trash, there were homeless people living under the bridge, it was not the kind of setting that you'd expect to see in a major urban park." The landscape designers, in particular Roger Green of Crosby Hanna and Associates and Ken Dockham, Director of Operations with Wascana Centre Authority, realized that the west end of the lake could be improved by placing fill there instead of hauling it east of the Trans-Canada Highway—and that the savings in hauling costs would partially pay for the improvements. Dockham was also largely responsible for the decision to turn the old Broad Street Bridge abutment into Pine Island.

The design also provided for a future pedestrian underpass under the Albert Street Bridge to connect

Above (left): The Wascana Lake rehabilitation project produced odd visual juxtapositions like this one of a shovel at work in front of the Saskatchewan Legislative Building. *(Photo by Edward Willett.)*

Above (right): A variety of equipment at work in Wascana Lake along the north shore, not far from the bandstand. *(Photo courtesy Wascana Centre Authority.)*

the Devonian pathway system and the Wascana and university pathway systems. "We didn't have the funds at the time to do it, but it can be accommodated now quite readily," Clifton said. Once in place, it will allow a cyclist or walker to travel from the northwest to southeast corners of the city while crossing only one street.

Other elements of the design, now in place, included an improved launching area for boats and canoes by the marina, and a significant amount of improved aquatic habitat. "There are designed shallows around Spruce Island and Willow Island as fish spawning and early rearing areas," Clifton said. "There are designed weed beds, if you can believe it. They're there on purpose to create habitat for fish and aquatic species. There are two major deep pool areas in the lake taken down to 7, 7.5 metres to provide overwintering areas for fish."

Oxygenation of the west lake's water is now aided by aeration systems, which include the highly visible fountain in front of the Legislative Building, the waterfall on Pine Island, and underwater bubblers.

Clifton is full of praise for the design team. "People could grasp a concept and go away and work on their portion of the design and come back and put it on the table and, my goodness, it fit with the other things people were working on. . . . The high quality of the designers . . . made the whole thing possible."

Environmental concerns force changes

Several modifications had to be made as the project advanced. Originally, the concept for the lake east of Broad Street was to use the spoil to build up Wascana Hill and fill in the old Goose Island area. "The concept . . . became unacceptable from an environmental perspective because that is part of a registered migratory bird sanctuary," Clifton explained. In the end, very little was done to the east lake beyond putting in berms to stabilize the north and east shores, which were eroding badly from wave action, and creating Science Island, a small island in front of the Saskatchewan Science Centre which will eventually be connected to shore by a walking bridge and will become an instructional venue. Wascana Hill was also redesigned and relandscaped.

The original plan was in place by July 1—and, said Clifton, "promptly blown out of the water." The regulatory authorities rejected it in mid-September as environmentally unacceptable. A complete redesign followed. "It was literally day and night by the design group," Clifton said. Fortunately, the regulators put in some overtime, too. "I think it was October 4 we actually got agreement in principle."

The project was tendered on October 9. Tenders closed in three weeks—right in the middle of a provincial election campaign. That forced another delay before the tenders were awarded. Dominion Construction of Regina was the contractor; Broda Construction of Prince Albert, a Saskatchewan company started by Cas Broda of Kamsack, was subcontracted to handle the deepening of the lake.

STOCK PHOTO

What lay beneath

Many people speculated what might be found on the lake's bottom. Bob Hughes says the contractors discovered "five bicycles, two bathtubs, one park bench, 27 beer bottles, 20 sandbags, one canoe, one Walkman, one radio and seven tires." (Hughes, p. 119)

A few more items turned up as excavation proceeded, but many sunken items probably never saw the light of day. "When you take the material out in big bites there could be things there that you never see," Clifton said. As well, snow came fairly early in the fall, creating a 15-cm blanket. "We don't know what was on the surface, because it was hidden by the snow until the very last thing in the spring. As we started excavating around Willow Island, the snow finally melted and we saw the surface."

Police officers' pre-excavation predictions finally came true: three firearms, presumably hurled into the lake from the overlook, turned up near the island. "We called the city police and they recovered them," Clifton said. "I'm sure they were examined with care." Among other unusual items they found: a roll of IMAX film, a filing cabinet, and "a University of Regina bag with books and identification from 1974."

The plug is pulled

Draining of the lake began on October 15. Most of the water escaped within four or five days: there wasn't much there. "We drained it for two days and we noticed the geese looked to be standing up in the middle of the lake. In fact, they were; in some parts of the middle of the lake there was less than half a metre of water," Clifton said. "It was quite amazing."

Broda Construction began building its camp and access roads near the end of November. The contract specified that all construction traffic should stay on the lake bottom. "To get the designed amount of fill out of the west lake required a loaded 50-ton truck to cross Broad Street every 90 seconds," Clifton said. "In fact, in some parts of the work it was down to less than 80 seconds. That amount of interference with surface traffic would have just been way too dangerous." Instead, Broda excavated truck underpasses beneath the Broad Street and Trans-Canada Highway bridges. At the same time, the contractor built crossings over major waterlines servicing south and southeast Regina.

The haul roads reached the west lake on December 22, and then operations were shut down until after the New Year . . . which began with a long, bitter cold spell. "When the contractor came back . . . it took him two days to get the equipment started and thawed out, because it was all hydraulic," Clifton recalled. "They had to warm all the hydraulic oil to get the equipment functioning."

Aside from that, those working on the dig welcomed the cold weather. "We were the only ones in Regina who wanted it to be 40 below for the whole winter," Clifton said. "Cold was good for us. The longer it stayed cold, the happier we were . . . The only way you could handle

STOCK PHOTO

Who put that dam there?

There are always surprises on a major project, Clifton said. One such surprise ran from what is now Pine Island to the mainland, following the original course of Broad Street.

"We found that the CPR had built a massive concrete dam that nobody remembered. It was completely hidden below about a foot and a half of muck," Clifton said. "About two weeks into the project, bang, we hit this concrete.

"It was a massive structure. It was built like a World War I bunker, massive, strong concrete, a lot of steel, sitting on timber piles—and the timber piles under the structure were as fresh as the day they were put in.

"That was a bit of a hiccup . . . but (with) the massive equipment the contractor had, he was able to attack it and within the week he was by it and going."

Right page: The trucks used in the deepening of Wascana Lake take a final drive out of the lake bottom along the haul road on March 22, 2004. *(Photo courtesy Wascana Centre Authority.)*

the upper 1 or 2 metres of muck was frozen solid. Otherwise you could literally pour it out of a pail."

The equipment ran 24 hours a day, seven days a week, except for four occasions when poor visibility halted work, three times due to snowstorms and once due to ice fog. Initially the work fell behind schedule, but the contractor adjusted, bringing in more equipment and picking up speed. In the end, Clifton said, "He actually met the design schedule pretty damn close."

That was partly because, as Clifton said, "Mother Nature co-operated," but most of the credit went to

Left: Willow Island, one of the features created during the first deepening of Wascana Lake in the 1930s. It and Spruce Island were the idea of provincial minister of public works Jimmy Bryant. Today Willow Island is a beautiful and popular picnic spot, but at the time it was derided as "Bryant's folly." *(Photo by Edward Willett.)*

Right page: Even after the lake had filled again, work continued on the project. Here a diver works on installing the aeration fountain located near the north shore of the lake. *(Photo courtesy Wascana Centre Authority.)*

those operating the machinery, who demonstrated both staying power and skill. "There was not a lost-time accident on the job. Even with that density of traffic and heavy vehicles, there were no collisions. There was one truck that slipped off and lay over on its side, but it was set back up and went right back to work."

The speed and scale of the operation were mind-boggling. The machinery included one of the largest excavators in the world, a 2004 CAT385 BL 8yd^3. With a bucket capacity of 6.1 cu. m, it could fill a giant dump truck with just three scoops. The next largest excavator had a bucket capacity of 5.35 cu. m. On average, it took only 45 seconds to fill an empty truck.

The end is near

Heavy equipment and thick, soft mud don't mix: everyone knew the spring thaw would end the digging. Ray Pentland, Wascana Centre Authority consulting hydrologist, monitored the snow pack and tried to predict when runoff would occur. His analysis indicated there was a 50/50 chance of runoff any time after March 10; after March 20 that went up to two out of three.

Sure enough, by mid-March it was no longer freezing. That made the final stages of work particularly tough. "The trucks were sinking up to their axles in the bottom of the lake," Clifton said. "They were using the biggest bulldozers they had to push the trucks out of the lake. Their fuel consumption in that final 10 days or two weeks almost doubled to what it was during the winter, because it was just so soft and took so much power."

Excavation ended on March 22. "We really worked to the last available day, and we needed the last available day," Clifton said. As the equipment was pulled out of the lake, he said, "the water was coming in. (The contractor) started pulling back the haul roads and doing some of the cleanup work in the east lake and about a week later, on Good Friday, the dike holding the water back in the east lake overtopped and the water came into the lake with a whoosh."

Landscaping, completing the walkway under Broad Street, completion of Pine Island and other tasks took the rest of the summer, but, said Clifton, "The time-sensitive work had been accomplished, the stuff that absolutely had to be done before the water started running was all done . . . the big-dollar work and the most obvious portions of the work were completed pretty much on time and pretty much right on budget, too."

The final statistics: the trucks made approximately 156,000 one-way trips, amounting to 1.012 million km (equivalent to driving to the moon and back, and then more than halfway to the moon once more). They removed 1 327 600 cu. m of dirt from the lake, an average of 19 000 cu. m per day. The lake's

STOCK PHOTO

water capacity, estimated at 1500 cu. dam (1,216 acre-feet) in 1990, is now estimated to be 2500 cu. dam (2,027 acre-feet). "It was really a heavy mining operation going on right in the heart of the city," Clifton said.

He concluded, "I think really the striking thing of this whole project is the integration of disciplines, and how each brings a contribution.

"So much of the engineering effort is not obvious to the public. What is obvious is a superb landscape design portion. The integration of the engineering and landscape design components (creates) a landscape that is both functional and pleasing to the public. You can see by the way the public flocks to the facilities now."

A hundred years from now, the weeds may encroach on Wascana Lake again. But for the next century at least, it should continue to be a shining jewel in the Queen City's crown, thanks to the efforts of Saskatchewan landscapers, contractors, workers . . . and engineers and geoscientists.

CHAPTER 8 SOURCES

Books

Hughes, Bob. *The Big Dig.* Regina, 2004. Leader-Post Ltd.

Macdonald, R.H. *Thorough: An Illustrated History of the College of Engineering, University of Saskatchewan, 1912-1983.* Saskatoon, 1982. The College of Engineering, University of Saskatchewan.

Websites

CBC News. "Saskatoon's 1st bridge closed for good." November 3, 2005. (February 3, 2006.) <http://www.cbc.ca/sask/story/bridge051103.html>

Online Learning Centre, Saskatoon Public Schools. *Saskatoon - The City of Bridges.* (February 3, 2006.) <http://olc.spsd.sk.ca/DE/resources/bridges/index.htm>

Video

Meewasin Valley Interpretive Centre. *Bridge Tales. Saskatoon - The City of Bridges.* Online Learning Centre, Saskatoon Public Schools. <http://olc.spsd.sk.ca/DE/resources/bridges/bridgetales.exe>

Personal interviews conducted by the author

Clifton, Wayne. President of Clifton Associates Ltd. December 2, 2005.

CHAPTER 9

Academia

A geological engineering class in session at the University of Saskatchewan. *(University of Saskatchewan Archives A-1597.)*

Students who took the first class ever held on the University of Saskatchewan campus, a gasoline engineering short course, pose in front of the Engineering Building in March 1912. *(University of Saskatchewan Archives A-1822.)*

Engineering and Geoscience at the University of Saskatchewan

The Early Years

In 1909, Walter Murray, president of the University of Saskatchewan (U of S) noted in his annual progress report that the Colleges of Arts and Science and Agriculture had been formed at the then two-year-old institution, and that within the College of Agriculture there was provision for a Department of Agricultural Engineering. He also announced the hiring of Professor A.R. Greig, a McGill University graduate in mechanical engineering, as professor of Agricultural Engineering and superintendent of buildings, making Professor Greig the first engineer on the U of S staff.

In July of that year Murray showed Greig the current assets of the university on the eastern bank of the South Saskatchewan River: grass, two buildings (a horse barn and a settler's shack being used as a granary), and a small clump of poplar saplings.

A fact-finding committee had recommended that the university adopt an architectural style called "Collegiate Gothic," previously used at Bryn Mawr and Princeton. The board of governors agreed, and chose the Montreal architects who had recently designed the medical building at McGill. The architects presented their plans in November 1909 for a number of campus buildings, including the Agricultural Laboratory (later the Engineering Building). Tenders were called for on March 21, 1910, and Smith Brothers and Wilson's tender of $792,291.85 was accepted.

Some of the buildings were to be made of stone; others, including the future Engineering Building, of red brick. While the architects can take credit for their beauty, historian Arthur Silver Morton credited Greig with "the excellence of their construction." (Macdonald, p. 3.)

A sod-turning ceremony was held on May 4, 1910, and Prime Minister Sir Wilfred Laurier laid the cornerstone for the Administration Building later that year. Supervised by Greig, 17 teams of horses and 16 labourers started excavations. By the fall of 1911 exterior work was almost complete.

"... born of mixed parentage in someone else's building, but ... a sturdy infant with high ideals and an expanding future."

Using university workers and others rather than contractors, Greig also oversaw construction of a second group of buildings. By 1912 the university's core buildings for the next several years were in place, for a total cost of $1.5 million.

Engineering classes begin

In March 1912 Professor Greig taught 174 students the university's first engineering class (and the first class of any sort to be taught on the university's permanent grounds), a short course on gasoline engines.

Agricultural engineering, Greig's specialty, was the only kind of engineering offered at first, and was so popular that it soon became one of the most important university departments.

In 1912 President Murray asked Chalmers Jack Mackenzie, a graduate of Dalhousie University in

Halifax and a partner in an Alberta engineering firm, if he would like "a job for the winter." (Macdonald, p. 20.) Mackenzie came to the university that fall as a sessional lecturer in civil engineering. He organized a course for his six students, taught descriptive geometry and drafting, and that spring held a survey camp.

The next fall he accepted a second invitation from Murray. During that year, 1913-14, a formal School of Engineering was formed within the College of Arts and Science. George H. Ling was named dean; Mackenzie was appointed professor of Civil Engineering. Feeling he needed more formal training, he headed to Harvard University for the 1914-15 school year to obtain a Master of Civil Engineering degree and returned to Saskatoon for 1915-16.

Murray's report for that year noted that "though it is not intended to attempt, for some years to come, more than the course in Civil Engineering, a student may complete one, and the greater part of two years in other Engineering courses at this University." (Macdonald, p. 22.) The courses available included railroad engineering, highway engineering, applied mechanics, structures, bridge design, foundations and masonry, engineering specifications and contracts, hydraulics, mechanical engineering, first-term shopwork (carpentry and joining) and second-term shopwork (smithwork and foundry practices).

The Faculty of Engineering met for the first time on October 20, 1915,

Left page: Boy Scouts and a band help celebrate the laying of the cornerstone for the University of Saskatchewan Engineering Building, ca. 1911. *(University of Saskatchewan Archives A-2991.)*

Above (left): Returned soldiers pose on a Case steamer inside the University of Saskatchewan Engineering Building in 1918. Over five years, almost 1,200 men attended agricultural engineering classes designed specifically for returned soldiers. *(University of Saskatchewan Archives A-2416.)*

Above (right): Class conditions in the early years of the U of S were rather primitive by today's standards. Here engineering students attend a lecture in the original Engineering Building, ca. 1920. *(University of Saskatchewan Archives A-400.)*

and the first engineering degrees (Bachelor of Science in Civil Engineering) were awarded in 1916 to Spencer Ball, George Oliver Thorn and Reuben John Haney.

Closed for the duration

And then the School of Engineering closed because the entire faculty and all of the students except for Haney, an American, had joined the Armed Forces. Fourteen University of Saskatchewan engineers served, and three, C.M. Bayne, J.K. Dawson and C.M. Harrington, were killed in action.

As soldiers, disabled or otherwise unfit for war service, began to return, the Department of Agricultural Engineering began to offer courses designed especially for returned soldiers, to 76 men in 1917 and 212 in 1918—both vocational courses (tractor engineering, motor mechanics, even motion picture projection) and degree courses. The special courses ended in 1922, but over five years almost 1,200 men enrolled in them.

Hard on the heels of the war came the 1918 influenza pandemic. Again, the school's activity almost ground to a halt. The School of Engineering didn't reopen until the fall of 1919, with 16 students in the first-year course and four in the fourth-year course. The faculty resumed meetings and began planning for a full four-year course and discussing possible additional courses in electrical and mechanical engineering. In 1921-22, the School of Engineering formally became the College of Engineering, with Professor Mackenzie as its first dean.

Those planning the university's growth had decided to focus on developing a few top-notch colleges and schools rather than spreading scarce resources over too many disciplines. One reason engineering was chosen to be developed was because its enrolment had surged from just 16 students in 1919 to 44 in 1921-22. Seven new staff members were hired.

At about the same time, the three Prairie universities decided amongst themselves that because of the development of hydroelectric power in Manitoba, clay in Saskatchewan and coal in Alberta, the University of Manitoba would focus on electrical engineering, the University of Saskatchewan on ceramic engineering and the University of Alberta on mining engineering. (Y.R. Anderson of Young, Saskatchewan became the first ceramic engineer to graduate from a Canadian university when he received his degree in 1925.)

Left page: Rows of forges, anvils and other equipment await agricultural engineering students inside the U of S Engineering Building in 1913. *(University of Saskatchewan Archives A-398.)*

Above: "We are! We are! We are the engineers!" U of S engineering students pose with surveying equipment in 1914. *(University of Saskatchewan Archives A-10841.)*

The fiery fate of the first Engineering Building

In 1917 the Engineering Building was extended to accommodate a tractor laboratory. In the early 1920s it received an additional wing to accommodate civil and hydraulic engineering. The new wing included a laboratory, three drafting rooms, three lecture rooms, faculty offices and locker rooms.

At about 2:30 a.m. on March 13, 1925, fire broke out in the tractor laboratory. The city fire station responded swiftly, but because the university was on the high side of the river, it had much lower water pressure than the city. A booster pump at the hydrant increased the pressure, but because the pump's capacity was greater than that of the supply pipe, the water soon ran out.

The fire gutted the building. Only the ceramics addition survived, ironically because of a fire door intended to protect the rest of the building from the risk of fire from the kilns.

The total loss was $267,000. Because the building was underinsured, funds set aside for a new Arts/Library Building had to be diverted to restoration of the Engineering Building. The new Arts Building had to wait three more decades.

The new, improved Engineering Building (with tile and concrete floors and walls instead of wood) went up in record time, and students were back at work there by January of 1926.

The change to a College of Engineering necessitated higher academic standards. Until the fall of 1924, only junior matriculation had been required for admission into the engineering program. After that, prospective engineers had to have either a full year of Arts and Science or senior matriculation.

With 14 graduates, the engineering class of 1925 was the largest yet. More significantly, unlike previous classes, almost all of the students were Saskatchewan-born and had started their courses in Saskatchewan as freshmen.

There would be many more changes over the years, but the College of Engineering, "born of mixed parentage in someone else's building" had indeed "proved to be a sturdy infant with high ideals and an expanding future." (Macdonald, p. 1.)

Birth of the Geology Department

Interest in geology in Saskatchewan increased when the western provinces received control of their mineral resources from the federal government in 1930. At the time, very little geological work had been done in the Precambrian rock of northern Saskatchewan, and the need for mapping and exploration of the resources there gave impetus to the formation of a Geology Department at the university.

Professor W.G. Worcester, an engineer, gave the first course in geology. Geologist F.H. Edmunds, a graduate of the University of Liverpool, joined the university staff in 1926 as an assistant professor in the Department of Soils. He gave his first class in geology in 1926-27. That fall, Dr. Justin S. DeLury came to the university. Together, he and Edmunds formed the first faculty of the new Department of Geology.

DeLury went back to the University of Manitoba the next year, but in 1929 Dr. J.B. Mawdsley, educated at McGill and Princeton, joined the staff. He had previously been employed by the Geological Survey of Canada, where he was involved in geological mapping in northwestern Quebec. For the next 11 years, Edmunds and Mawdsley were the faculty of the Department of Geology. A 1959 history of the department notes, "On account of staff limitations the apportionment of classes at times did not wholly correspond with their chief interests. However, Edmunds was far more at home teaching a class in mineralogy than Mawdsley was teaching a class in paleontology, for which, by necessity, he developed a great respect." (*The Concentrates*, p. 2.)

The Geology Department's laboratories and offices were located in the southeast corner of the Engineering Building. The first course in geology had been taught by an engineer, and the connection between the two departments remained strong: in 1935, the course in geological engineering began with the help of Dr. C.J. Mackenzie. "It resembled closest a mining engineering course with a strong geological option," notes the department history, and was designed "to meet the needs of students in the College of Engineering who wished to enter the mineral industry." (*The Concentrates*, p. 3.)

Left page: The original U of S Engineering Building in 1921. Just four years later the building was completely gutted by fire. *(University of Saskatchewan Archives A-1822.)*

Above: Dr. F.H. Edmunds, the first faculty member of the Department of Geology at the University of Saskatchewan and department head from 1961 to 1965. *(University of Saskatchewan Archives A-3290.)*

Above (left): Dr. J.B. Mawdsley, who with Dr. Edmunds formed the entire faculty of the U of S Department of Geology for its first 11 years. *(University of Saskatchewan Archives INS-225.)*

Above (right): Dr. C.J. MacKenzie, first dean of the College of Engineering at the U of S, speaks at a 50th anniversary engineering reunion on October 5, 1962. *(University of Saskatchewan Archives A-1453.)*

Right page: Dr. Amy Veawab, an Associate Professor of Environmental Systems Engineering at the University of Regina, conducts research in a U of R lab. *(Audio Visual Services, University of Regina.)*

With summer employment non-existent during the 1930s due to the depressed economy, 15 of the university's geology students formed parties in 1934 and spent the summer in the North. They gained even more experience during the next year, when the federal government supplied the Geological Survey of Canada with $1 million, primarily to give summer employment to students. More than 130 U of S students were part of field parties, including the 25 or so geology students.

The outbreak of the Second World War had a devastating impact on enrolment. The federal government wanted universities to train students for the war effort. Because geologists and mining engineers weren't considered particularly valuable to that effort, the number of geology students dropped so precipitously that in 1945 only one graduated.

All that changed with the end of the war. Returning veterans flooded the campus, and by 1948-49 the senior class consisted of 33 students, all geological engineers.

The early focus of the Department of Geology, reflecting the interests of its faculty, was very much on geological engineering—practical geology focused on finding and extracting minerals. Mawdsley and A.R. Byers (a McGill graduate hired in 1940) were both very active in exploring northern Saskatchewan. Dr. Glen Caldwell, head of the Geology Department in the 1980s, said of Mawdsley, "I think his first love was what I call mining geology . . . He was tremendously biased, I thought, in favour of geological engineering and had very little time for honours students in the straight science program . . . He used to refer to honours students as 'one-track Charlies', and claimed they had their head in the clouds rather than their feet on the floor, whereas he felt the geological engineers had their feet on the floor and could turn their hands to anything." (Kupsch and Hanson, p. 15.)

Caldwell had a different opinion. The prime role of the Geology Department, he said, is "providing students with a thorough grounding and a good understanding of the concepts and principles of all the major fields of the geological sciences." (Kupsch and Hanson, p. 17.)

The Department of Geological Sciences of the University of Saskatchewan strives to do that to this day.

"It's been an exciting time."

Engineering and Geoscience at the University of Regina

In 1910, the General Council of the Methodist Church of Canada appointed a board of governors to found Regina College. With help from the Massey estate in Toronto, and money raised within the province, the new school opened in 1911 with 26 students in temporary quarters. A new main building opened in 1912.

The two provincial schools weren't destined to remain completely separate for long. In 1925 Regina College became a junior affiliated college of the University of Saskatchewan, and in 1934, unable to pay its bills, agreed to a complete takeover by the university.

Teaching engineering in Regina

Returning Second World War veterans boosted enrolment at both Regina College and the University of Saskatchewan. (For example, engineering enrolment at the U of S was 1,156 in 1946-47, a record for many years.)

To reduce the load on the university, new courses were added in Regina: beginning in 1945-46, engineering students regularly spent their first year at Regina College. Then, in 1959, the U of S upgraded Regina College to a full degree-granting second campus, called the University of Saskatchewan, Regina Campus. Construction of new buildings began at the current site of the University of Regina.

The academic program of the new campus began to take shape. Engineering, the third professional college, was established in 1965.

In 1968 John Mantle, head of Mechanical Engineering at the University of Saskatchewan, was named the first Dean of Engineering for Regina Campus. Key faculty members were hired to teach courses in mechanical, electrical, chemical and civil engineering.

But almost before it really got going, the Faculty of Engineering faced a crisis. In 1971 the newly elected New Democratic Party government called for a review of engineering education. Bruce Cooke, who would become the school's second dean, wrote, "It was rumoured that the cost of renovations to the College of Engineering in Saskatoon and the cost of a new Physics/Engineering building in Regina triggered the review. There were also questions being asked such as 'How many engineers do we need in Saskatchewan?' and 'Why are we duplicating programs when there are so few engineering students?'" (Cooke.)

Time to close it down?

In March 1972 the government cancelled a planned engineering building at Regina Campus, and the board of governors appointed a commission to study the engineering programs in both cities. The commission chair, Dr. Philip Lapp, was a consulting engineer from Ontario who had chaired a review of engineering education in that province in 1966. While awaiting the outcome of that study, the board of governors decided to continue to offer the first two years of engineering in Regina and to proceed as planned to introduce a third year in the fall of 1972.

The Lapp Report, issued in December 1972, recommended that all engineering instruction take place in Saskatoon and that the Regina program be closed down after the 1973-74 academic year. Instead, the board of governors directed the Faculty of Engineering to continue with the two-year program for three to five years, and in the meantime to investigate "alternative developments in the field of applied science and engineering technology for Regina Campus." (Pitsula, in Knuttila, p. 25.)

Many of the faculty members John Mantle had attracted to Regina Campus weren't interested in teaching in a feeder program for the University of Saskatchewan. A majority left.

For those who remained, things began looking up with the act of the legislature that established the

University of Regina (U of R) as an institution separate from the University of Saskatchewan as of July 1, 1974.

"It was a real challenge to make it go"

That same year Art Opseth, who had studied under Mantle in Saskatoon, joined the faculty. "They needed some people to help keep it going," he explained, adding, "Some people couldn't understand why I would leave what they called a secure government job (doing computer programming and systems work) to work for anything as uncertain as engineering at the U of R . . . It was a real challenge to make it go."

Left page: The first four Deans of Engineering at the University of Regina, photographed in 1994. From left to right, John Mantle, Bruce Cooke, Sesh Seshadri and Don Stewart. *(Photo courtesy Art Opseth.)*

Left: U of R students prep the race car they designed and built for the Formula SAE (Society of Automotive Engineers) competition in Pontiac, Michigan, in May of 1997. Derrick Mann is at the wheel; Michael Kuntz is on the left and Mervin Armstrong on the right. *(Photo courtesy Art Opseth.)*

With between 100 and 150 students, the faculty began to push for the establishment of a full engineering program. A study to determine which programs could be offered that wouldn't duplicate those at the University of Saskatchewan recommended programs in the areas of industrial, regional and electronics engineering.

That study also recommended that the University of Regina focus on "systems" engineering. Thus, regional engineering would emphasize large-scale civil engineering projects such as water systems and transportation systems; industrial engineering would turn out engineers capable of being the only engineer of a small company or the project leader for a large company; electronics engineers would be concerned with a whole range of digital applications.

"When you look at an engineering project you have to take a look at everything that's there, and that includes people, resources, technology, environment—everything," is how Opseth put it. "The idea was that you need two kinds of engineers: you need the specialist, and you also need the person who can take the broad approach."

In 1979, John Mantle described systems engineering this way: "A systems engineer is an engineer trained for the analysis, design, and evaluation of natural and man-made systems with a particular awareness of the economic, humanitarian, environmental and administrative implications of his/her work." (Opseth paper.)

The study also recommended that U of R students be required to participate in a co-operative education program, so that they learned on the job as well as in the classroom.

Four conditions

The new engineering programs were approved by the Universities Commission, subject to certain conditions."With very few faculty members, detailed design of the new programs was a problem," Bruce Cooke wrote. "An appropriate and fortunate solution was found. The programs were designed using

Co-operative education at the U of R

The co-operative education program at the U of R was the first such program in Western Canada, and just the third or fourth in the whole country. "It is something I've been particularly proud of," Art Opseth said.

In the co-op program today, students study at the university for three semesters, then work with a company for one four-month term. After that, they alternate work and study terms, thus graduating with four work terms or 16 months of practical experience.

Finding companies to take the co-op students has always been a challenge, Opseth said, though it was easier when the University of Regina was the only Western Canadian university offering co-operative education. Companies from BC to Ontario have taken on U of R students.

The Physics, Computer Science and Chemistry Departments and the Faculty of Administration also offer co-op education, Opseth said. "It started in Engineering, then it spread out through the others." He said it's been a great success because of the practical experience it offers students. "It's helped a lot of them make a decision about their careers. Many students after their first work term have decided they're going to switch what their major was, because they ended up in a job they really liked."

There's another major benefit: "Students who have been with the co-op program all the way . . . have made enough money to actually pay their way through university, compared to coming out and having a big student loan to pay off."

As well, Opseth said, a lot of co-op students, when they come back for their last academic year, already have job offers. "The employers find these young people and like them; they want to hang on to them."

Industrial Advisory Committees. This provided a focus on relevance that greatly aided the success of the programs that were created, and the Industrial Advisory Committees have continued to provide valuable assistance and advice since their inception." (Cooke.)

The lack of funding made it important to use existing courses outside the Faculty of Engineering (i.e., math, physics, chemistry) as part of the program, something that has continued.

The first third-year class started in January of 1979. Many students still chose to transfer to the University of Saskatchewan after the first two years, so the first graduating class under the new programs, in 1981, consisted of only six students. Efforts in the 1980s focused on increasing the number of graduates. Meanwhile, the struggle to explain the "unconventional" programs and justify the very existence of engineering at the U of R continued.

Today the University of Regina boasts about 750 engineering students (the University of Saskatchewan has around 1,500). There are 300 graduate students. The university offers programs in Industrial Systems Engineering, Electronic Systems, Petroleum Systems, Environmental Systems and Software Systems. There still isn't an engineering building; engineering students take classes in the Education Building, where offices and some research labs are also located. Other research has expanded to some of the new buildings, including the Petroleum Technology Research Centre, TRLabs, and more.

Opseth, who served as acting dean for one year and assistant dean for 13 or 14 years, was in charge of co-op programs for a while, became associate dean of special projects before he retired, and continues to be a sessional lecturer, sums up the development of the Faculty of Engineering at the University of Regina very simply:

"It's been an exciting time."

Left: A student in the Environmental Systems Engineering program at the University of Regina works in the lab. *(Audio Visual Services, University of Regina.)*

Geoscience at the U of R

Sessional lecturer Dr. H. Lyn Jones, a research geologist with the Saskatchewan Department of Mineral Resources, taught the first introductory geology class at the University of Saskatchewan, Regina Campus, in 1965. The following spring the first faculty member, Dr. W.A. (Tony) Gordon, who had been teaching at Oberlin College in Ohio, was hired; he taught introductory geology classes during the 1966-67 and 1967-68 terms.

In 1966 a Geology Advisory Committee began exploring the possibility of establishing geology as a discipline within the Division of Natural Sciences. Early in 1967 it finalized the requirements for a

Left: Students at work in one of the U of R's Petroleum Systems Engineering laboratories in the Petroleum Technology Research Centre. *(Photo courtesy Audio Visual Services, University of Regina.)*

Right page: A 2200-m-deep well is drilled on the U of R campus in December of 1979 as part of an investigation, led by Dr. Laurence Vigrass, into the feasibility of using geothermal energy to heat buildings on the Saskatchewan prairie. *(Photo courtesy Audio Visual Services, University of Regina.)*

The U of R geothermal well

For about 10 years in the 1970s and 1980s, Dr. Laurence Vigrass was the chief investigator on a project designed to demonstrate the use of geothermal energy from sedimentary basins to heat buildings.

In December of 1979 and January of 1980, the team drilled a well 2200 m deep on the south side of the U of R campus. "The temperature at the bottom of the well is about 62 degrees Celsius, and we could produce that water at about 60 degrees Celsius," Vigrass said. "One calculation we made was that we probably could have supplied about 3 megawatts of heat energy, not as electrical energy, of course, but energy for heating purposes."

The plan was to use the geothermal energy to heat a field house or athletic complex. The hot brine from deep underground wouldn't have circulated directly through the building; instead, it would have heated fresh water via a heat exchanger, and the heated fresh water would then have circulated the heat through the facility. That way, Vigrass said, "just in case there is a breakage of some sort then at least you're not dealing with that really salty water."

Lloyd Barber, who was then president of the U of R, was very supportive of the idea, but unfortunately, Vigrass said, the demonstration facility was never built. "The well is still there, but we never did drill the second well you would need to dispose of those really salty waters," he said.

But even though the project never reached the demonstration stage, Vigrass's research was instrumental in development of the Temple Gardens Mineral Spa in Moose Jaw. There, warm (45 degrees C), mineral-laden water from 1350 m underground is piped to the surface for the enjoyment and therapeutic benefit of residents and tourists alike.

three-year ordinary bachelor's degree, which the Division of Natural Sciences approved in principle in May. Courses included introductory geology, mineralogy, paleontology, geomorphology, structural geology, petrology and stratigraphy. (An additional course, sedimentology, was approved but not taught until 1972.) A second faculty member, Dr. John Lewry, was hired in 1968. A sessional lecturer, Dr. D.M. Kent, also taught some courses.

Dr. Laurence Vigrass joined the department staff in November of 1968, and began teaching structural geology in the 1969 winter semester.

The first three students to obtain B.A. degrees in geology from the U of R graduated in the spring of 1969 and went on to the University of Saskatchewan to obtain B.Sc. (Advanced) degrees.

Dr. Geoff Parslow joined the faculty that same year, and during 1969-70 enrolment in the program jumped to between 15 and 21 students. The curriculum for a four-year B.Sc. degree program was approved in the spring of 1970, and when former sessional lecturer Dr. D.M. Kent became the fifth faculty member in September of 1971, the M.Sc. program was accredited.

In the fall of 1972, geology field camps began. Over the next nine years, field camps alternated among the Little Rocky Mountains in Montana, Star Lake in Manitoba, and the Red Lodge-Yellowstone areas of Montana. (After that, they moved to the Lundbreck Falls area in the Crowsnest Pass and to the Flin Flon area in Manitoba.)

Vigrass, who became head of the Geology Department in 1988 and retired in 1991, called field work "the heart of geology." At the field camps, students practise "geological mapping — for example, description of stratigraphic sections, from older rocks up to younger rocks. It's a very important part of geological training."

Until the summer of 1973, when the department moved into College West, faculty offices were scattered around the Lab Building and undergraduate laboratory facilities were in the Classroom Building. Research labs were located at the College Avenue campus; they didn't move for another three years.

Over the ensuing years the department has continued to grow in enrolment, number of faculty, and number and variety of courses offered.

Every university department develops its own special areas of interest that grow primarily out of the interests of the faculty members. The U of R Geology Department's foci include Precambrian geology ("We have and have had quite a number of Precambrian geologists," Vigrass said) and, on what he called the "soft rock" side, "they have done mostly carbonate geology, which is limestones and dolomites."

"We did quite a lot in the 1980s, and especially in connection with geothermal energy," said Vigrass. (See sidebar.) "That study of fluids has been continued. They have a faculty member now that takes a rather different approach, but still does a lot of work with fluids, both free fluids—fluids that flow—and those that are enclosed in minerals and in inclusions in minerals. It's pretty interesting, but quite specialized, too.

"I think one of the changes that really has occurred since I was hired is, when I came there, environmental geology was something you might devote a few minutes to in your last lecture of the year in a geology class. Now they direct that course toward environmental geology, and that's things like disposal of all materials, both solid wastes and liquid wastes. That's also matters of pollution and environmental hazards, including landslides and earthquakes and volcanoes.

"It has always been a university that has had quite a strong leaning toward the applied side of geology."

The exterior of the Canadian Light Source in Saskatoon. *(Photo courtesy Canadian Light Source Inc., University of Saskatchewan.)*

"It speaks to the calibre of Saskatchewan engineers that this kind of project was designed in province . . ."

The Canadian Light Source

The University of Saskatchewan now boasts a light source a million times brighter than the sun—but don't worry, that doesn't mean the province's population will soon have to be issued industrial-strength sunscreen and welders' masks. All you can actually see when you're standing next to the Canadian Light Source is an impressively large open space filled with cables, pipes, and more pipes and more cables, laid out in a large circle—or two circles, really, one inside the other.

The Canadian Light Source is a synchrotron, a device that spins electrons in a circle, and in the process gives off powerful light in a range of useful wavelengths. It's one of the most advanced in the world. It's also one of the most impressive engineering achievements in Saskatchewan history.

How it works and what it does

Out of sight down in the basement of the Canadian Light Source, an electron gun injects electrons at a relatively low energy into a stainless steel tube that contains . . . well, nothing; it's full (if that's the right word) of a vacuum emptier than the space around the International Space Station. Those electrons are accelerated by microwaves down a linear accelerator into the inner "booster ring," where they're accelerated even more, to almost the speed of light, approximately 300 000 kilometres per second (kps). Finally they're sent into the outer "storage ring."

Just as a car taking a corner too fast emits a chirp of sound and a puff of rubber smoke, electrons forced around a curve at high speed emit photons (particles of light) over a wide range of wavelengths, from infrared all the way up to hard X-rays.

Right page: How a synchrotron works. *(Graphic courtesy Canadian Light Source Inc., University of Saskatchewan.)*

A little energy, in the form of radio frequency waves, is constantly fed into the storage ring to replace that which is radiated away as light. As a result, at lower energy levels, electrons can keep circulating in the storage ring for up to 50 hours—by which time they have travelled further than from Earth to Pluto.

As the photons leave the synchrotron, they immediately begin moving in a straight line. This means they can be drawn off in "beamlines" and put to work by researchers needing a powerful beam of light for their studies.

Bright light can help you see small objects more clearly. But you can only see extremely small structures if the wavelength of light striking them is even smaller than they are. Visible light can illuminate objects as small as bacteria, but to see anything smaller, you need shorter wavelengths—soft and hard X-rays.

Each beamline, then, is first sent through an "optics hutch," where unwanted wavelengths are filtered out. The beam of the desired wavelength is passed on to an "experimental hutch," where it illuminates the study sample. Data collected in seconds may then take weeks or months to analyze.

The Canadian Light Source's brilliant illumination can be used to examine the ways living cells interact with drugs, to learn more about (and possibly treat) cancer, to discover better ways to clean up pollutants, to improve materials (from motor oils to plastics), and even to make nanodevices (nano- meaning "extremely small," just a few atoms in size).

Some of the most exciting work is in the field of protein analysis. Proteins, from which all life is made, have enormously complex structures, and very few have been analyzed in detail. Understanding the structure of certain key proteins may help us develop new drugs, understand diseases, improve crops and more.

The history

Scientists first began calling for a Canadian synchrotron in the 1970s, and the University of Western Ontario proposed one back in 1974 (the preliminary design for it, coincidentally, was drafted by University of Saskatchewan mathematician Roger Servranckx).

Nothing came of that proposal, however, and for another quarter century Canada was the only major industrial nation without a synchrotron or plans to build one. That forced Canadian researchers who needed synchrotron light to travel to other countries.

Then, in 1990, the Canadian Institute for Synchrotron Radiation (CISR) was founded. Its president was Michael Bancroft, who with Bill McGowan had spearheaded the Western proposal in 1974. The organization sought to coordinate and facilitate Canadian synchrotron-based research and improve access to synchrotron facilities.

In 1994 the Natural Sciences and Engineering Research Council (NSERC) Committee for Materials Research Facilities recommended that Canada move toward constructing its own synchrotron

Seeing Is Understanding

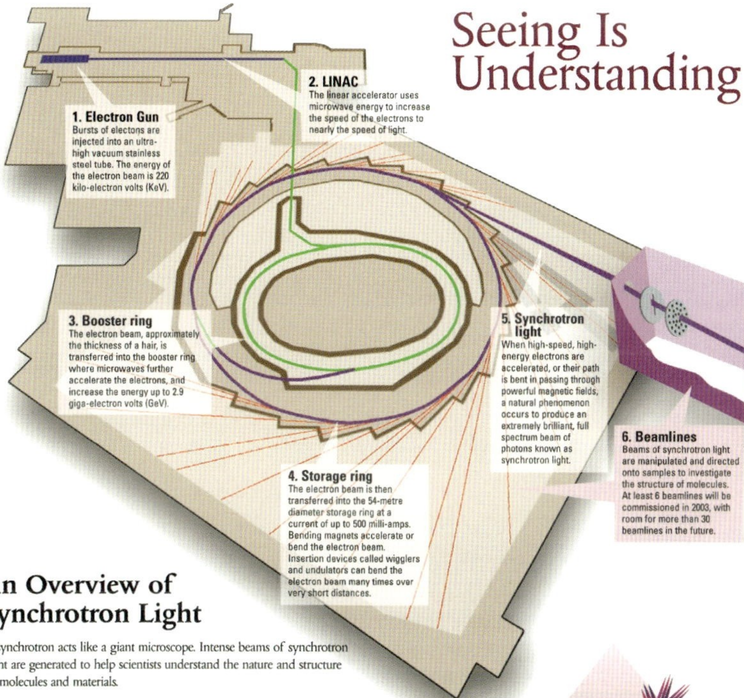

Synchrotron Analysis of Molecules

Beams of synchrotron light are transferred into the beamline 'mini-laboratories' where the chemical analysis takes place. For any given chemical question or molecular analysis, selected wavelengths of synchrotron light and a variety of synchrotron techniques are necessary to characterize different aspects of the nature and structure of molecules or materials. A few days of data collection can result in several months of data interpretation.

Beamlines can be categorized into groups or types based on selected wavelengths. These groupings include infrared, soft X-ray, and hard X-ray beamlines.

5
www.lightsource.ca

An Overview of Beamline Operations

"One thing I have learned in a long life: that all our science, measured against reality, is primitive and childlike and yet it is the most precious thing we have."

Albert Einstein

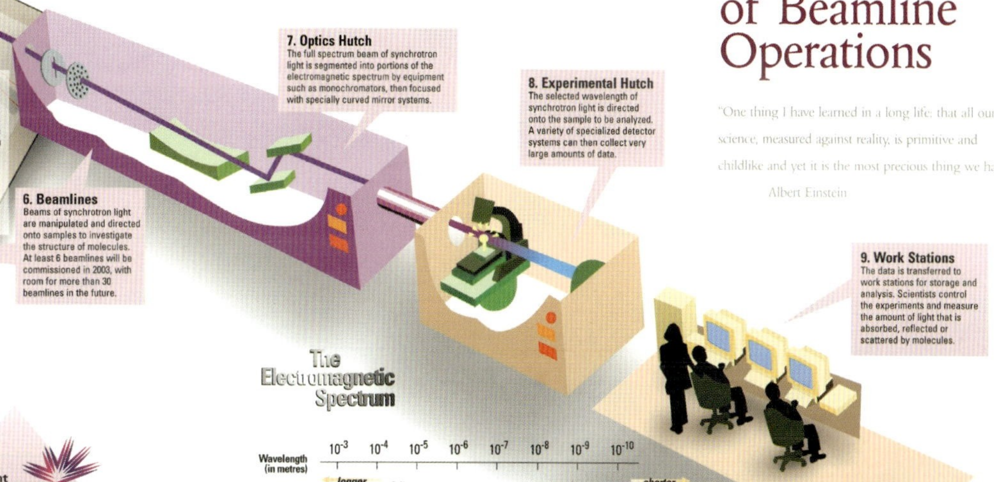

An Overview of Synchrotron Light

A synchrotron acts like a giant microscope. Intense beams of synchrotron light are generated to help scientists understand the nature and structure of molecules and materials.

The Canadian Light Source will be one of the most powerful in the world, producing synchrotron light that is millions of times brighter than sunlight. This new research tool at the University of Saskatchewan will help to find solutions to challenges in health, the environment, and advanced materials.

What is light and what does it do?

Light is energy that travels in waves. Visible light is the small portion of the electromagnetic spectrum that can be seen with human eyes. The spectrum is composed of varying energies of light that have different wavelengths.

The length of the wave of light, or wavelength, is very important to the scientist. To examine a material using some synchrotron techniques, the wavelength must be the same size or smaller than the material being examined. Some wavelengths are like specialized X-rays that can 'see through' some materials.

The data produced when synchrotron light interacts with various materials is analyzed to produce images and charts that can illustrate such things as the relationships of atoms within molecules.

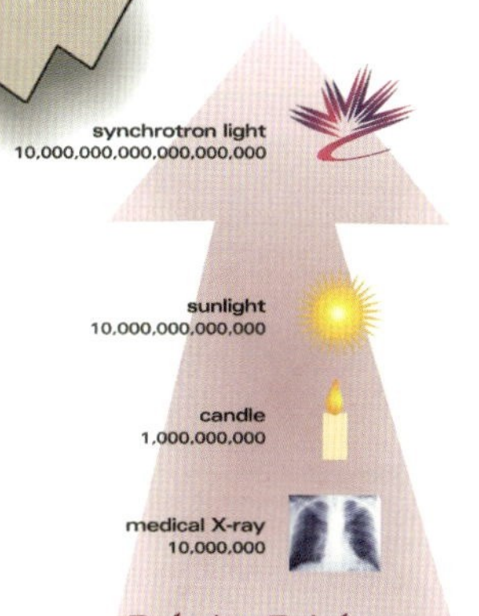

Relative Brightness

Photons per second per mm²
(comparisons at frequency of peak brightness)

The Electromagnetic Spectrum

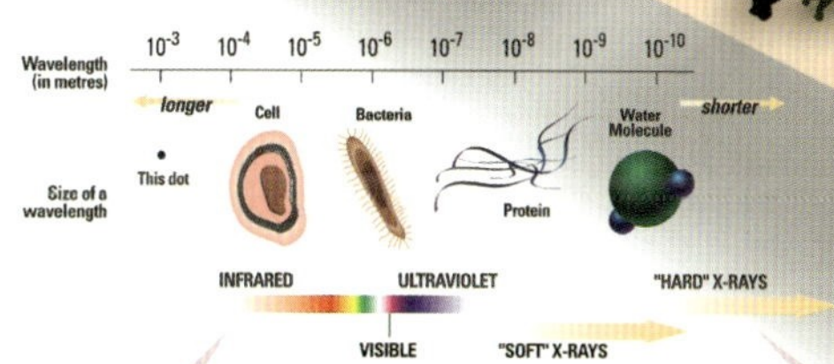

Infrared beamlines

Infrared (IR) beamlines will capture the wavelengths of synchrotron light that are longer than visible light. IR beamlines will explore the spectral resolution of gas phase molecules, the molecular vibrations of biological molecules, and the transition states of molecules under pressure in advanced materials.

Soft X-ray beamlines

Soft X-rays include wavelengths that are shorter than visible light. Soft X-ray beamlines will be used to study the chemistry and structure of gases, liquids and solids by measuring the absorption of the light, as well as the energies and directions of various particles such as electrons emitted after a soft X-ray photon is absorbed. A variety of soft X-ray microscopes will be available to perform chemical analysis of solids and surfaces at spatial scales of 20 nanometres (1 nanometre is 1 billionth of a metre).

Hard X-ray beamlines

Hard X-rays probe matter with light of very short wavelengths, about the same size as an atom. Synchrotron techniques on hard X-ray beamlines include crystallography, scattering, spectroscopy and microanalysis. Measurements can be made of the diffraction, or bending and scattering, of the synchrotron light as it interacts with sample materials. The protein crystallography beamline will determine the structure of biological macromolecules, the function of proteins, and the molecular interactions of potential pharmaceuticals to develop improved drugs.

facility. That year and the next, workshops were held to consider a possible design by the Saskatchewan Accelerator Laboratory, a branch of the Department of Physics and Engineering Physics at the U of S. At a meeting in May 1995 CISR decided to circulate the design to the Canadian scientific community and solicit bids. At that same meeting, the proposed facility received a name: the Canadian Light Source.

In 1996 NSERC's international peer review panel received and reviewed bids from the University of Western Ontario and the University of Saskatchewan. After studying the proposals and visiting both sites, the panel unanimously chose the Saskatchewan proposal and recommended that construction funding be made available as soon as possible.

Teams at the two universities then joined forces to produce a combined, enhanced proposal, which NSERC unanimously accepted on October 21, 1996. The endorsement by Canada's largest scientific agency prompted further funding. Primary sources of funds were the Canada Foundation for Innovation, the Canadian government, the Saskatchewan government, the Ontario Innovation Trust, the Alberta government, the University of Saskatchewan (which wholly owns the finished facility), the City of Saskatoon and SaskPower.

The main synchrotron facility cost $140 million, out of a total budget of $173.5 million.

The engineering

Right from the conceptual stage, the engineering firm most closely associated with the Canadian Light Source has been UMA Group Ltd. Founded in Saskatoon in 1911 and now based in Vancouver, UMA (formerly Underwood McLellan and Associates) has long been involved with projects at the University of Saskatchewan, including construction of the university's linear accelerator. (That's a device that accelerates subatomic particles in a straight line, then smashes them into a target: in the resulting debris, scientists find clues about the nature of matter.)

According to Mark de Jong, Director of Operations for the Canadian Light Source, engineers faced three main challenges in constructing the synchrotron facility: vibration, temperature control, and vacuum.

Researchers require that the beamlines from the synchrotron move from side to side by no more than 1 micron—just a fraction of any wavelength of visible light. "We constantly measure the position of the beam to within a micron," he said. "We do well, but it's not perfect, so we have a feedback system to correct that."

Vibrations too tiny to be felt are huge on that scale. Changes in temperature cause components to shrink or expand, the amount varying depending on the materials involved. Even the magnetic field that keeps the electrons circling the storage ring would fluctuate if the temperature changed, because the iron in the magnets would change permeability.

To minimize vibration, de Jong said, "The entire floor of the main ring is isolated from the main support structure." The perfectly smooth and level concrete main floor of the enormous open room (about 75 m by 75 m) in which the synchrotron is housed is separated from the outside walls by 2 cm of silicone, and supported by 720 piles driven up to 20 m deep.

Bob Gander, a University of Saskatchewan engineering professor, was one of the judges for an award of excellence presented by the Consulting Engineers of Saskatchewan to UMA Group Ltd. for its work on the synchrotron. He commented, "The most amazing constraint to me was that the floor has to stay level to 35 microns. It's phenomenal that they can design a floor which is a slab the size of a football field and can stay that stable." (CLS News Release.)

The temperature of the booster and storage ring enclosures has to be kept at 27 degrees C, and can vary

---

Left page: A fish-eye view of the Canadian Light Source. *(Photo courtesy Canadian Light Source Inc., University of Saskatchewan.)*

by no more than 0.1 degree. That stringent requirement led to the cooling system and pipes costing about four times the original estimate, de Jong said.

There are two types of temperature control, added Terry Johnson, Engineering and Technical Services Manager for the Canadian Light Source. Keeping the air flow through the tunnels at a constant temperature is the main stabilizing element, but the components are also actively cooled with deionized water.

The insulation and temperature control work so well that "you can have the whole north side in glass and not worry about extra heating costs," said Ray Richards, UMA's Deputy Project Manager. (During the design phase, researchers made natural light one of their top requests.)

Creating and maintaining the very hard vacuum within the synchrotron was also a challenge, de Jong

Left: Interior view featuring the Storage Ring, Infrared labs and the SGM-PGM Beamlines. *(Photo courtesy Canadian Light Source Inc., University of Saskatchewan.)*

said. "Compared to other places, it's particularly good. These facilities are unique in terms of vacuum."

Creating such a hard vacuum required careful attention to fabrication techniques, materials selection, welding techniques, cleaning and handling procedures and assembly procedures, said Johnson. Then the device had to be evacuated using several kinds of pump, and various "bake-out techniques" employed, because some materials release gases after being installed.

CLS innovations

The Canadian Light Source was state of the art as it was being built, and features several innovations. In order to keep the electrons circulating after they're injected into the storage ring, a little additional energy has to be supplied in the form of radio frequency (RF) waves. At the CLS the device that does this uses superconducting components—which means they have to be cooled by liquid nitrogen at -195.8 degrees C (the CLS uses about 20 000 l per month) and, in some cases, liquid helium at -268.9 degrees C (the CLS uses about 2000 l of that in a closed system).

Several synchrotron facilities around the world have also begun using superconducting RF devices since the CLS was built; many sent people to Saskatoon to see how the CLS does it.

The CLS also features a lot of computer-related innovations, including what was, when it was installed, the biggest VOIP (Voice Over Internet Protocol) system in Western Canada, and an optical fibre network. "There was a heavy emphasis on getting very high-speed data infrastructure," de Jong noted, not surprising considering the amount of data collected by researchers.

There are hundreds of computers, many of them embedded in various systems, and almost all running open source, free software. "Of all the synchrotrons, we have probably paid the least for our control software," de Jong said. There's such a high degree of automation that at night the synchrotron can be run by a single person.

"... the floor has to stay level to 35 microns. It's phenomenal that they can design a floor which ... can stay that stable."

Oh, and all the monitors use *Star Trek* fonts.

Looking down the road

Looking ahead, de Jong said, "Our challenge will be, 'What will the users want in 10 to 15 years? Will they still want us?'"

He said the cost of completely gutting and replacing the ring would be "pretty small," but a bigger challenge may be pushing the technology to substantially improve the synchrotron while keeping it the same size. "We are probably the smallest in this size in this energy level," de Jong said. On the other hand, that also means the Canadian Light Source has "one of the highest efficiencies around."

It's also one of the most cost-efficient synchrotrons ever built. "If the Department of Energy (in the United States) had built it, it would have cost twice as much," de Jong said. "And it's been done technically with a group that's half to a third the size of anywhere else."

Not only that, added Johnson, "A very large percentage of people within the engineering group are from Saskatchewan."

Bob Gander summed it up best. "It speaks to the calibre of Saskatchewan engineers that this kind of project was designed in province using local engineers," he said. (CLS News Release.)

What else can you call it but a brilliant success?

CHAPTER 9 SOURCES

Books

Dalzell, Matthew, ed. *Canadian Light Source Activity Report 2001 – 2004.* Saskatoon, 2005. University of Saskatchewan Press.

Knuttila, Dr. K. Murray, ed. *Heritage and Hope: The University of Regina into the 21st Century.* Regina, 2004. Canadian Plains Research Center.

Kupsch, W.O. and Hanson, S.D., ed. *Gold and Other Stories as told to Barry Richard.* Regina, 1986. Saskatchewan Mining Association.

Macdonald, R.H. *Thorough: An Illustrated History of the College of Engineering, University of Saskatchewan, 1912-1983.* Saskatoon, 1982. The College of Engineering, University of Saskatchewan.

Websites

CLS – Canadian Light Source. (April 13, 2006.) <http://www.lightsource.ca/>

Deo et Patriae: Events in the History of the University of Saskatchewan. (April 13, 2006.) <http://scaa.usask.ca/gallery/uofs_events/>.

Other sources

The Concentrates. Saskatoon, 1959. Ore Gangue. (Special issue of student geology group newsletter marking the Golden Jubilee of the U of S.)

Cooke, Bruce. Manuscript titled "Notes on the establishment of engineering education in Regina."

"Engineering Firm Wins Award for Work on CLS." Canadian Light Source News Release. June 21, 2001.

Information supplement on the Canadian Light Source. *The Globe and Mail.* Friday, October 22, 2001.

Kent, Don. Untitled manuscript on the history of the University of Regina Geology Department.

Personal interviews conducted by the author

Opseth, Art. Former assistant dean, Faculty of Engineering, University of Regina. November 17, 2005.

Vigrass, Laurence. Former head, Department of Geology, University of Regina. April 10, 2006.

Group interview with Mark de Jong, Director of Operations for the Canadian Light Source, Ray Richards, UMA's Deputy Project Manager for the CLS, Terry Johnson, Engineering and Technical Services Manager for the CLS, and others, August 15, 2005.

STOCK PHOTO

CHAPTER 10

Made in Saskatchewan

Top: Installing a virtual weigh station that uses International Road Dynamics technology. *(Photo courtesy International Road Dynamics.)* **Bottom:** Today's new oriented strand board plants, like this one in Meadow Lake, are direct descendants of the Wizewood Waferboard Plant in Hudson Bay, the first viable waferboard plant in the world. *(Photo courtesy of Saskatchewan Industry and Resources.)*

> “We’re celebrating our 40th anniversary . . . and in this business, it’s a real accomplishment to survive and do well for that long.”

Above (left): SED Systems technicians install a payload into a National Research Council sounding rocket.

Above (right): Manning telemetry equipment at the Churchill Rocket Research Range in Churchill, Manitoba. More than 65 SED-designed payloads were launched by sounding rockets from the Churchill range over a 22-year period beginning in 1967.

(Photos courtesy SED Systems.)

SED Systems

It’s not where you’d expect the tale to start, but in a way SED Systems (that’s pronounced S-E-D, by the way, not “said”) grew out of the Great Depression.

During the 1930s Saskatchewan’s economic situation became so dire that all unmarried professors at the University of Saskatchewan had to take one or two years of unpaid leave to help the university balance its budget. Balfour W. Currie and Frank T. Davies of the Physics Department chose to take their unpaid leave in 1932-33, the Second International Polar Year, when researchers around the world were measuring the Earth’s magnetic field and observing auroras and the weather to help them address problems in navigation, radio communication and weather forecasting.

Currie, 30, and Davies, 28, spent the year in the Far North, at Chesterfield Inlet—and thus began space and upper atmospheric research at the University of Saskatchewan.

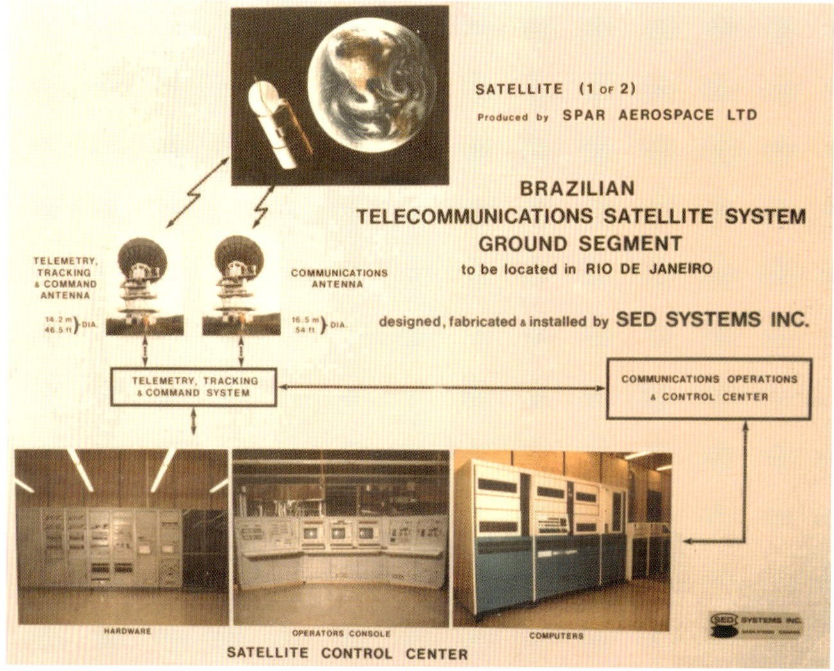

Over the next two decades, university researchers would probe the aurora borealis with radar, weather balloons and rockets. And in 1957, Currie formed the first space and atmospheric science institute in Canada, the Institute of Space and Atmospheric Studies (ISAS).

The birth of SED

In the 1960s, Canada had an extensive sounding rocket launch program, carried out by the National Research Council (NRC) of Canada. Sounding rockets (like the famous Canadian-built "Black Brant") carry a payload above the atmosphere, to the edge of space, but don't put it into orbit. Instead they return to Earth after a flight of about half an hour.

Until 1965, although ISAS designed various experiments and instruments for sounding rocket launches, NRC carried out most of the payload development and engineering itself. But in 1965, when the NRC decided "to contract out some of those services, Dr. Alex

Above (left): Technicians in Brazil monitor communications satellites using equipment designed and provided by SED Systems.

Above (right): In the 1980s SED provided the complete ground segment for a commercial communications satellite program in Brazil, as detailed in this poster. The contract helped make SED a world leader in satellite communication and telemetry.

(Photos courtesy SED Systems.)

“The company today is quite different from the one formed back in 1965.”

Above: An SED engineer tests a component of the Canadarm during its development in the 1980s.

Right page (top): SED provided four of these radio frequency subsystems to Inmarsat for its new I4 satellite network.

Right page (bottom): The Decimator is a new computer card SED developed for low-cost spectrum measurement and analysis.

(Photos courtesy SED Systems.)

Kavadas saw an opportunity for a commercial venture,” and the Space Engineering Division (SED) of ISAS was born. It employed one engineer, three technicians, and Dr. Kavadas, the director. (SED Systems.)

The first rocket payload designed by SED was launched from the Churchill, Manitoba rocket range in 1967. Over the next 22 years, SED built and launched more than 65 rocket payloads. (SED Systems.)

But SED quickly moved beyond building payloads to building space-related technology for use on the ground, most of it communications oriented. It was the beginning of a long-term change. “The company today is quite different from the one formed back in 1965,” said Don Epp, marketing director for SED Systems. “We started out building primarily space flight instrumentation, apparatus that actually flew in space for primarily scientific applications. The customers at that time were almost exclusively Canadian government clients. Today our core business is satellite communications, very much on the commercial side of things. Most of our customers are private companies. Most of those customers are not in Canada, they're in the United States or Europe primarily, and everything we build these days stays on the ground.”

SED converted the Prince Albert Radar Laboratory into a tracking station for remote-sensing satellites. It operated that station for the next 16 years, and, in fact, received the first images of Earth taken by the first Landsat satellite. (SED Systems.) Soon after converting the Prince Albert facility, SED delivered the ground control system for Canada's communications technology satellite (CTS), launched on January 17, 1976.

SED goes private

By the early '70s SED's many contracts had made it financially independent. Industry competitors complained, however, that it had an unfair advantage because it was a division of a university-based research institute. Partly for that reason, in 1972 it became a private, limited liability company, owned by the university, and changed its name to SED Systems.

Innovations poured out of SED through the '70s. Not all were space-related. Its agricultural

Late in the '80s SED Systems Inc. moved into its current facilities at 18 Innovation Boulevard, but its pace never slackened.

instrumentation division, which eventually became a separate company (one of more than a dozen with roots in SED) developed grain-loss and seeder monitors and a mastitis detector for dairy cattle. SED also won the Governor-General's award for design for the High Line Data Acquisition (HILDA) system, to measure vibration frequencies in high-voltage power lines. (SED Systems.)

Toward the end of the decade, SED installed the most northerly known geosynchronous satellite communications ground system at Eureka, which lies at 81 degrees N., began work with Hughes Aircraft Company to provide satellite test and ground control equipment (a relationship that continues today with Hughes's successor, Boeing Satellite Systems), and worked with Telesat Canada to provide satellite tracking systems in Canada and Guam. (SED Systems.) In May of 1976 it left the university campus for new facilities near the Saskatoon Airport.

In the early '80s, SED Systems, which by then employed more than 200 people and had annual sales of $12 million, reorganized into SED Systems Inc. (SED Systems.) In 1986, Fleet Aerospace Corporation, based in Ontario, acquired controlling interest.

Brazil "really got us going"

During the 1980s SED signed a contract to provide the complete ground segment for a commercial communications satellite program in Brazil. "That really got us going, not only in developing the systems and the technology involved, but also in terms of the project management and capability to bring together large projects like that, particularly for a foreign customer, where we needed to build the systems and take them to site, install them and make sure they were working," said Epp.

For the Brazilian program SED developed an in-orbit test system which can verify that a communications satellite is operating properly. During the 1990s SED established itself as an international leader in this field, delivering in-orbit

Above: A technician works at a custom test bench set up in SED's manufacturing area.

Right page: The 35-m antenna SED built for the European Space Agency in Cebreros, Spain. The ESA uses the antenna to communicate with deep-space probes.

(Photos courtesy SED Systems.)

testing systems to satellite companies around the world. "Now we've taken some of our test capability that we developed in space applications and we're applying it to terrestrial communications markets," Epp said. "We've established a small office in Calgary that focuses on those sorts of activities."

Over the years the company had built up some excess capacity in its manufacturing, assembly and test capabilities, Epp said, and so in the '80s it began building high-reliability, high-performance, high-quality, low-volume, difficult-to-build electronics systems to customers' designs. Much of that work is associated with military, surveillance or security applications. SED entered the field by building radar displays for Transport Canada's modernization of its air-traffic control network.

SED also provided communications equipment for Canada's new patrol frigates, one of its largest contracts ever. But space remained at the core of the company's business.

On to Innovation Boulevard

Late in the '80s SED Systems Inc. moved into its current facilities at 18 Innovation Boulevard, but its pace never slackened. SMS, a supra-thermal mass spectrometer, became the first foreign instrument to fly on a Japanese satellite, *Akebono*. SED began developing ground-based equipment to test Canadarm II, the robotic manipulator that is now installed on the International Space Station. (SED Systems.)

It also developed ground systems and test equipment for RADARSAT-1, Canada's first remote-sensing satellite—and provided operational support for it, which continues today. "We fly RADARSAT-1 from Canadian Space Agency (CSA) facilities in St. Hubert, but we also have a telemetry tracking and command station located at our facility here in Saskatoon," Epp said. "We recently received a new contract to continue those services." Since RADARSAT-1, launched in November of 1995, was originally intended to have five years of operational life, SED has managed to keep it flying more than twice as long as expected. "The satellite has worked very well and provided imagery of features, events and conditions, not only in Canada but globally, that help in environmental modelling, mapping, forestry and a range of agricultural applications," Epp said.

SED also flies the CSA's SCISAT, a smaller satellite studying the depletion of the ozone layer over Canada and the Arctic.

In 1990, SED was purchased by Calian Technologies of Kanata, Ontario. In 2000, it became a division of Calian Ltd.

Digital satellite radio became available to Canadian listeners in 2005. SED has been involved in this new satellite application for more than six years, providing both WorldSpace and XM Satellite Radio with uplink systems that allow them to send radio content up to their satellites for rebroadcast to listeners.

Giant antenna, giant contract

One of the biggest achievements of the 1990s was signing a $23-million contract—the largest ever awarded to a Canadian company by the European Space Agency (ESA)—to build a 35-m deep-space antenna system. SED has now built two.

The first was built in Australia. The second, completed in 2005, is in Spain. Both are roughly 12 storeys tall. The main antenna reflector of each structure is basically "600 tonnes of moving steel," Epp said, which can rotate 360 degrees and be elevated 90 degrees. Each massive dish is within 0.5 mm of being perfectly smooth, and can be pointed with a precision of several thousandths of a degree.

"The antenna tracks deep-space probes heading to the outer reaches of the solar system, and sends commands to activate and control the on-board systems, as well as receiving information and data," Epp said.

One probe the European Space Agency is currently communicating with is *Rosetta*. By mid-November of 2005 *Rosetta* was already 297 million km from Earth, so distant that a signal travelling at the speed of light (approximately 300 000 km per second) was taking 16 minutes and 29 seconds to reach Earth. The satellite will rendezvous with comet 67P/Churyumov-Gerasimenko in the summer of 2014, release a lander to make the first controlled touchdown on a comet nucleus, and then fly alongside the comet as it heads toward the inner solar system, to observe how the frozen comet changes as it is warmed by the sun. (ESA.)

Another spacecraft the antennas communicate with is *Mars Express*, which since December of 2003 has transmitted images and other data of unprecedented detail from Mars orbit.

The massive antennas were joint efforts by several companies, with SED as the prime contractor. "We have subcontractors all over Europe," Epp said. "Our biggest subcontractor is based in Germany. They look after the mechanical portion—they do some of the work in their facilities and build up the basic bits and pieces, which are then shipped to site. The mechanical structure doesn't really come together until it is assembled on site.

Above: The first 35-m deep-space antenna SED built for the European Space Agency, in New Norcia, Australia. *(Photo courtesy European Space Agency.)*

Right page: Intelligent toll roads using International Road Dynamics systems, like this one in San Pedro Sula, Honduras, allow motorists to pay without having to stop. *(Photo courtesy International Road Dynamics.)*

"There are always new applications in satellite technology."

"We look after the radio frequency components, the transmitters, the receivers, frequency converters, all those sorts of things. We would typically bring the components into our facility and integrate them—make sure they're working as a package. Then all of that is shipped off to the site and then integrated into the antenna." From start to finish, production of the Australian antenna took about four years.

The Australian antenna is at about 120 degrees east longitude. The one in Spain is close to the prime meridian. The ESA has tentative plans for a third one somewhere at about 120 degrees west longitude. That would make it possible for the ESA to communicate with its deep-space probes any time of day, no matter where the Earth is in its rotation. (Epp.)

Pluses and minuses

With customers all over the world, there are pluses and minuses to being located in Saskatchewan, Epp said. "The downside is that most of our customers aren't here—there are essentially no customers here. So our proximity to customers is an issue, but I think it's fair to say it would be regardless of where we were located . . . our customers are spread all over the world. We spend a lot of time in airplanes travelling back and forth.

"The upside of being here is we have a good standard of living but a relatively inexpensive one. We're in competition with companies in California or the eastern United States or various parts of Europe, and the cost of our labour can be kept somewhat lower than if we were operating out of some other area." SED Systems currently has about 230 employees, including about 76 engineers and several more who have an engineering background but are working in management or marketing.

The company also finds its proximity to the university is a plus. "We still have a close working relationship with the university, and most of the engineers we hire are University of Saskatchewan grads," Epp said. "Our business is obviously specialized enough that we need to do a lot of on-the-job training and our people need to gain experience through hands-on projects, but both the engineering and technical schools provide us with good candidates with strong basic skills and credentials.

"We're celebrating our 40th anniversary this year, and in this business to survive and do well for that long is a real accomplishment. We've seen a lot of companies come and go. It's continuing to be exciting work. There are always new applications in satellite technology, and continuing opportunities for us to be involved."

International Road Dynamics

In 1950 a young man named Arthur Bergan joined the Saskatchewan Department of Highways as an instrument man on road survey crews. In 1961 he obtained a bachelor's degree in civil engineering from the University of Saskatchewan. He continued to build roads, designing and supervising the construction of highways in northern Saskatchewan.

In 1964 Bergan earned his master's degree and started teaching at the University of Saskatchewan in the Department of Civil Engineering. He also began research into vehicle accidents and safety. In 1972 he earned his Ph.D. in civil engineering from the University of California at Berkeley, where his studies focused on pavement management and how highway loads impact pavement life.

In the mid-'70s, Dr. Bergan helped establish a network of transportation research centres, including one at the University of Saskatchewan (for which he was

Above: Dr. Arthur Bergan, who founded International Road Dynamics to commercialize the weigh-in-motion scale he invented in the 1970s.

Right page (left): Workers install an IRD virtual weigh station near Saskatoon.

Right page (right): IRD's smart information signs only turn on when needed to warn drivers of problems ahead, and can provide more detailed information than ordinary signs.

(Photos courtesy International Road Dynamics.)

"Most people agree we are the leader worldwide."

coordinator) under the mandate of Transport Canada. Shortly thereafter he headed up a research project which led to the development of a weigh-in-motion scale. The scale proved so successful that in 1980 Dr. Bergan formed a company, International Road Dynamics (IRD), to commercialize it.

Today IRD employs more than 140 people and has annual sales of more than $30 million.

No need to stop

Trucks operate under weight restrictions because if they are too heavy, they damage roads and can be unsafe. Traditionally, truck weights have been checked at roadside weigh stations. Trucks pull into the weigh station and, sometimes after having spent a considerable amount of time in a queue, drive onto a large scale.

Weigh-in-motion scales weigh trucks while they're travelling at highway speed. As trucks pass over the device, it automatically identifies those that might be oversize or overweight so they can be called in to a weigh station for further inspection.

Dr. Bergan's weigh-in-motion scale launched IRD, but it took a series of additional innovations to propel the Saskatoon-based company to the position of a world leader in what are now called "intelligent transportation systems."

Dr. Bergan's son Terry is the President and Chief Executive Officer of International Road Dynamics today. (Dr. Bergan is still chairman of the board of directors.) As important as the weigh-in-motion scale itself is the interface that converts the mechanical force measured by the scale into a digital signal for transmission to a computer. "You've got thousandths of a second to weigh something that's moving at highway speeds," Terry Bergan said. "As technology moved forward you went from a scale which had a platform that was 2 feet (wide) . . . to now we weigh with a sensor which is a couple of inches (wide). The next part of the innovation is the algorithm, the interpretation of that signal, putting it in useful form."

There are both horizontal and vertical forces at work, plus many variables, including roadway materials (Portland cement, say, versus flexible pavement), the external temperature (which influences the roadway's rigidity), how much the vehicle is vibrating or bouncing, even how moist the subgrade is. The computer program has to "wade through all these various factors that can affect your accuracy and ability to weigh a vehicle," Bergan said.

Putting the data to use

But intelligent transportation systems do more than just collect vehicle weights. "New camera technologies have allowed us to read licence plates," Terry Bergan said, "so we can use the information that the scale provides and integrate it into a national registration system and look at all of the records of vehicles."

As he pointed out, "the majority of trucks that are on the highways are properly registered, are safe and are not overweight. The innovation that has happened over the last five years as we've installed more and more of these systems is that we can keep the majority of the trucks on the highway—they never have to report to a scale."

Besides weighing trucks, weigh-station staff conduct safety inspections and check credentials and licensing. "With databanks you know which trucks have been recently inspected, you know which truck fleets typically have a safe record, and so you're not penalizing the good corporate citizens any more. You're targeting the ones that don't obey the law,"

Bergan said. "That saves time and money." It's also safer, he noted, because "you don't have trucks changing lanes any more trying to report to weigh stations." Not surprisingly, he said, "The American Trucking Association and Canadian Trucking Association are strong supporters of ours."

In Saskatchewan, large, heavy trucks deliver grain to elevators. IRD partnered with the Saskatchewan government to create the Automated Truck (Location) Monitoring System, the first of its kind in North America. It monitors the compliance and movement of specially permitted trucks (such as certain oversized grain-hauling trucks) that are allowed to operate on some secondary roads provided they use "road-friendly" equipment and do so during off-peak hours. IRD installs onboard computers, Global Positioning Satellites and cellular communications devices in the trucks, then monitors their movements and compiles a monthly report.

"You take that technology and you install it throughout the country," Terry Bergan said, "and from an engineering point of view you've got really good solid data of road usage, understanding pavement damage, safety aspects of highway transportation, because you know information on every vehicle that's on the highway."

Traffic management tricks

The technology can also be used for traffic management. Information from roadway sensors and cameras can be combined with data from environmental sensors, and used to update automated signs warning drivers about potential problems.

One example: "smart" work zones. "When a driver is approaching a construction area, when he sees the signs, he also sees that there's congestion ahead," Bergan explained. "The amount of time it takes to get through a work zone is going to depend on how long the line-up is . . . Smart signs and signals will provide that information in real time. As well, the signs only turn on if they're needed."

Studies have shown that drivers slow down when they first see a work sign, but speed up again if they don't see any congestion. "Many accidents happen because drivers see signs and don't see the danger and so they accelerate again," Bergan said. "Then they come up over a hill and all of a sudden they're coming up on car lights or cars that are stopped . . . If you have smart signs, the only ones that are turned on are the ones in the vicinity of the congestion, so they see the congestion, they stop."

Safety is the focus of several of IRD's products. Oregon's Green Light Project made extensive use of IRD systems and software. One example: systems that advise truck drivers of the safest speed for negotiating a downhill slope, based on the angle of the descent and the truck's weight.

International, but firmly Saskatchewanian

Although IRD's head office is in Saskatoon, it has a United States corporate office in Spring Grove, Illinois, 19 sales and service locations across the United States, a wholly owned subsidiary in Santiago, Chile, a joint venture in Sao Paulo, Brazil, a sales office in Bogota, Colombia, a wholly owned subsidiary in Delhi, India, and a joint venture in China.

Left page (top): Workers install a virtual weigh station in a highway near Saskatoon.

Left page (bottom): A truck rolls through the International Road Dynamics-designed virtual weigh station near Saskatoon.

Above (left and right): Inside IRD's manufacturing facility in Saskatoon.

(Photos courtesy International Road Dynamics.)

Above: Dan Bonnet, systems engineer for International Road Dynamics. *(Photo courtesy International Road Dynamics.)*

Right page: Logs surround the Wizewood Waferboard Plant in Hudson Bay in 1961. *(Saskatchewan Archives Board R-B5678 -11.)*

In many countries, Terry Bergan said, there are huge pressures on how highways are financed. "Governments just don't have the money to finance highways or maintain them, and so . . . toll roads." Traditional toll roads involve drivers stopping at booths to pay their toll. Intelligent toll roads obviate the need for booths—or stopping. "You can use the technologies that we have and automate toll roads so people pay with automatic debit cards. Trucks pay by weight, without stopping." One such road using IRD systems is the Fredericton-Moncton highway in New Brunswick.

Despite its international activity, IRD remains firmly ensconced in Saskatoon. "We do the majority of engineering and manufacturing here," Bergan said. There are about 30 people in the engineering and research staff in Saskatoon, but there is engineering expertise throughout the company.

That's not surprising, since "we started off as a bunch of engineering types who had great ideas, who somehow had to figure out how to run a company," said Brian Taylor, then senior director of technical solutions with IRD, in an interview with the Saskatoon *StarPhoenix* in 2000. He pointed to company founder Dr. Art Bergan as the inspiration for the company's ongoing creativity. "Dr. Bergan was recognized throughout North America for his expertise. As his career went on, he gained worldwide recognition. I came out of his graduate program. He afforded many opportunities to his students, and we took advantage of those opportunities and built on that." (*StarPhoenix*.)

"We hire a lot of Saskatchewan engineers," Terry Bergan said. "The University of Saskatchewan is one of the main reasons we're here. SIAST (the Saskatchewan Institute of Applied Science and Technology) is also a very important business partner for us for technical staff. Overall, we're very pleased with the people we're able to attract locally."

He also pointed to the positive relationships the company has built with local suppliers and subcontractors. "We focus on being the technical innovators. As a result, we subcontract out certain parts of the manufacturing process, such as machining, the fabrication of metal enclosures or the production of printed circuit boards. We've been very fortunate in being able to source out the majority of these contracts to excellent firms in Saskatoon. There's only a very small percentage of our equipment that is manufactured outside the city, and then it will very likely be produced nearby in Western Canada."

Today International Road Dynamics has an installed base of more than 2,500 systems in North, South and Central America, Australia, Asia and Europe.

Dr. Art Bergan, who started it all, was inducted into the Saskatchewan Transportation Hall of Fame on October 17, 2001.

The World's First Viable Waferboard Plant

On August 12, 1959, the *Hudson Bay Post Review* wrote, "The announcement by Wizewood Products Limited, Prince Albert, to build a $2,500,000 plant for the production of poplar building board in Hudson Bay, promises a tremendous boost to the forest product industry in this area."

It also marked an engineering first for the province and the world: the Wizewood plant became the world's first viable waferboard manufacturing plant.

Canada's first plywood plant opened in 1913, and Hudson Bay's first plant in 1948. But waferboard was something new. In 1954, at meetings of the Forest Products Research Society, J.d'A. Clark and A.L. Mottet described structural panels made from thin wafers of wood glued together. (Koch and Springate, p. 160.) By 1958 a waferboard plant using western red cedar was in operation in Idaho—but the timing wasn't right, and the plant's products failed to catch on.

STOCK PHOTO

"Pioneers in an undertaking unique on this continent"

The first made-in-Saskatchewan waferboard

The Wizewood waferboard plant began production in Hudson Bay in September 1961, under general manager J. Korodi. On September 27 the *Hudson Bay Post Review* reported that the first 4-by-16 waferboard slid off the production line at 11:30 a.m. on September 23. The newspaper described the process: "After several mechanical delays, the 'pad' of poplar wafers mixed with dry resin and wax moved slowly into the giant forming press. Here the pad was compressed into a board 5/8 of an inch thick under a pressure of 200 pounds per square inch. Heated to a temperature of 380 degrees the wax and resin melted to provide a bond for a wooden wafer . . . The test waferboard proved highly satisfactory as the wafers had bonded very securely and the resinous finish presented an attractive appearance. The stability of the 4 by 16 panel is superior to similar wood products."

A few weeks later the paper reported that waferboard inventor Dr. J.d'A. Clark attended an "informal meeting and smoker" at the plant on November 1. Dr. Clark told the more than 60 plant employees present that they were "pioneers in an undertaking unique on this continent."

Dr. Clark believed that as the forests' larger trees were felled, plywood manufacturers would need substitute products. He said waferboard provided a ready solution, and on the Prairies also provided a means of utilizing what was otherwise regarded as a "weed

Right page: A worker inspects a newly formed panel at the Wizewood Waferboard Plant in Hudson Bay in 1961. *(Saskatchewan Archives Board R-B5678 -3.)*

CASE
W S
CASE

tree." Dr. Clark also noted that the Wizewood plant had been very well engineered, and, said the newspaper, "hinted at a limitless need for the product as time went on." (*Hudson Bay Post Review*, November 8, 1961.)

During 1962, production problems were sorted out and demand for "Aspenite" grew. On January 9, 1963, a fire caused more than $1 million in damage, but rebuilding began at once, and the plant resumed operation well before its second anniversary.

Sold to MacMillan Bloedel

Still, said Peter Koch and Norman C. Springate in the March 1983 *Journal of Forestry*, the Wizewood plant, like the Idaho plant before it, had "marketing problems," and so it was sold to MacMillan Bloedel on April 14, 1965. "Under this management," they wrote, "waferboard found its way into the structural sheathing market in competition with West Coast plywood. Success was due mainly to its freight advantage." (Koch and Springate, p. 160.)

According to Koch and Springate, waferboard on the Prairies was first used in the construction and remodelling of farm buildings and fences and as a general utility panel, applications in which it proved itself satisfactory. It later received National Building Code approval for use in the construction of homes and two-storey apartment buildings.

Demand increased so much that in January 1968, MacMillan Bloedel announced a multi-million-dollar expansion to more than double the plant's capacity, making it the largest particleboard complex in Canada and eventually bringing total employment in the plant and surrounding woods to around 260.

Another major upgrade, worth $14.8 million, took place in 1983. The original press, "old No. 1," was shut down at 8:49 a.m. on August 24 of that year. During its 22-year lifetime, the press produced more than 1 million 4-by-16 sheets of Aspenite, equivalent to 11,000 carloads of building material.

By that time, the technology that would displace Aspenite and other waferboards was already in the works. Called oriented strandboard (OSB), its panels are created from long and narrow flakes, or strands, formed into mats with three or five layers. The strands in each layer are oriented perpendicular to the strands in adjoining layers, giving OSB superior strength and flexibility compared to ordinary waferboard, whose wafers are randomly distributed in the panel.

By 1989, most of the original waferboard plants had converted to oriented waferboard, a kind of halfway step between waferboard and OSB. By the time Weyerhaeuser bought Macmillan Bloedel in 1999, the Hudson Bay plant was producing OSB. Weyerhaeuser initially kept the plant (called an OSB 1000 plant) in operation while it built a brand-new "OSB 2000" plant capable of producing higher-grade OSB. But it wasn't long before there was an "indefinite market-related shutdown" (Saskatchewan Rate Review Panel) of the OSB 1000 plant, a shutdown that later became permanent.

The new OSB 2000 plant, though, is the original waferboard plant's direct descendant. All the other OSB plants around the world, including the newest one in Saskatchewan, at Meadow Lake, are at least first cousins.

"OSB products will improve," predicts an article published on the Building Materials and Wood Technology website operated by the Department of Natural Resources Conservation at the University of Massachusetts. Author Paul Fisette notes, "OSB manufacturers can formulate their process to provide virtually any property they want. They can build panels to resist high relative humidity, deliver more strength, or provide a harder surface . . . One thing is certain: OSB is in our future."

And that future started right here in Saskatchewan.

Left page: Dry wafers are loaded into the feeder to be turned into Aspenite at the Wizewood Waferboard Plant in Hudson Bay in 1961. *(Saskatchewan Archives Board R-B5678 -5.)*

CHAPTER 10 SOURCES

Books

Macdonald, R.H. *Thorough: An Illustrated History of the College of Engineering, University of Saskatchewan, 1912-1983*. Saskatoon, 1982. The College of Engineering, University of Saskatchewan.

Newspaper and magazine articles

Koch, Peter and Springate, Norman C. “Hardwood Structural Flakeboard – Development of the Industry in North America.” *Journal of Forestry*. March 1983. pp. 160-161. < http://www.srs.fs.usda.gov/pubs/4701/392.pdf>

Saskatoon *StarPhoenix*. "International Road Dynamics world leader in Intelligent Transportation Systems." December 13, 2000.

Worobec, Dale. "International Road Dynamics: A leader in the Intelligent Transportation Systems Industry." *Commerce & Industry*. October/November 2001. pp. 71-75.

Other publications

Saskatchewan Rate Review Panel. *Report to the Minister of Crown Investments Corporation on: The Proposal from SaskPower for Changes in Electrical Rates, March 23, 2001.* <http://www.saskratereview.ca/common/pdfs/saskpower_mar_23_01.pdf> (March 6, 2006.)

Websites

European Space Agency (ESA). *Rosetta*. (February 3, 2006.) <http://www.esa.int/SPECIALS/Rosetta/index.html>

Fisette, Paul. “Choosing Between Oriented Strandboard and Plywood.” *Building Materials and Wood Technology,* University of Massachusetts Amherst. (March 3, 2006.) <http://www.umass.edu/bmatwt/publications/articles/osb_vs_plywood.html>

Saskatchewan Industry and Resources. "SED Systems Antenna Listens to the Galaxy." *25 Saskatchewan Science Achievements*. (February 3, 2006.) <http://www.sk25.ca/Default.aspx?DN=92,88,16,1,Documents>

SED Systems. *Company History*. (February 3, 2006.) <http://www.sedsystems.ca/AboutUs/Company/History/index.php>

Town of Hudson Bay. *Local History – Newspaper Articles on Hudson’s Bay Forest Industry.* (March 3, 2006.) < http://www.townofhudsonbay.com/local_history/articles.html>

University of Saskatchewan Archives. *Balfour W. Currie - The Second International Polar Year.* (Feb. 3, 2006.) <http://scaa.usask.ca/gallery/northern/currie/>

Personal interviews conducted by the author

Bergan, Terry. President, International Road Dynamics. November 25, 2005.

Epp, Don. Marketing director, SED Systems. November 18, 2005.

CHAPTER 11

Establishing the Professions

Above and next page: Engineers gather in Regina on October 29, 1938, for a banquet to celebrate the signing of a co-operative agreement between the Engineering Institute of Canada and the Association of Professional Engineers of Saskatchewan. *(APEGS archival photo.)*

The first sentence of the first chapter of this book states, "Engineers—and engineering projects—were few and far between on the Prairies in the years before Saskatchewan's birth as a province."

But as the rest of the book makes clear, in later years engineers abounded. As the province boomed, cities, bridges, roads, buildings, sewers, telephone systems, power stations, dams and more sprang up everywhere: and designing and building all those things were engineers.

They were practising professional engineering, but they weren't professional engineers—not as the phrase is now understood. *The Engineering and Geoscience Professions Act* defines the "practice of professional engineering" as "any act of planning, designing, composing, measuring, evaluating, inspecting, advising, reporting, directing or supervising, or managing any of the foregoing, that requires the application of engineering principles and that concerns the safeguarding of life, health, property, economic interests, the public interest or the environment."

But it defines a professional engineer as "an individual who is registered with the association as a professional engineer," and when Saskatchewan began, there wasn't an association to register with.

There is now: The Association of Professional Engineers and Geoscientists of Saskatchewan (APEGS). And 2005, which marked the province's centennial, marked APEGS's 75th anniversary.

APEGS's roots are in another organization, born in 1887 as the Canadian Society of Civil Engineers ("civil" as in "not military," to distinguish its members from military engineers), whose object was "to facilitate the acquirement and interchange of professional knowledge among its members." (The Engineering Institute of Canada.) The society was headquartered in Montreal, but regional branches began to form as early as 1890.

The Regina branch formed in 1915. In 1918, it expanded to become the Saskatchewan branch.

As it moved into the 20th century, the Canadian Society of Civil Engineers, feeling the word "civil" in its name was too restrictive, formed four engineering sections: general, electrical, mechanical and mining. In 1918 an act of Parliament removed "civil" from the name entirely, changing the organization's name to the Engineering Institute of Canada (EIC).

The pursuit of professional status

In August 1918, at the Second Professional Meeting of the EIC in Saskatoon, a hot topic of discussion was legislation regarding the professional status of engineers, then being suggested in Manitoba, Saskatchewan and Alberta. Members felt it important that the EIC promote uniform legislation in all provinces.

At the next Annual General Meeting in Ottawa in February 1919, the EIC formed a committee to draft a model act. The result was endorsed by the EIC's membership and council and passed on to the various branches and provincial divisions. In 1920 acts based on the EIC's model were proclaimed in British Columbia, Quebec, Manitoba, Alberta, New Brunswick and Nova Scotia. Within three years the remaining provinces followed suit—except for Prince Edward Island and Saskatchewan.

"In advance of the times"

In Saskatchewan, the model bill was submitted to the provincial government, which returned it with suggested changes. Stewart Young wrote in 1962, "Well do I remember the then Acting Premier, Hon. C. A. Dunning, saying, 'Gentlemen, I fear that you are in advance of the times.'" (Young, "The Saskatchewan Branch, E.I.C.") He was right: it took several more years, until May 1, 1930, for Saskatchewan's first *Engineering Profession Act* to be passed.

Reginald Bing-Wo served as registrar of the Association of Professional Engineers of Saskatchewan (APES) on a part-time basis from 1952 to 1967, then full-time from 1967 to 1985. He said one reason it took so long for Saskatchewan to pass legislation was that "in those early days there was a lot of objection to having such an act, because a lot of the 'engineers' that were practising and working at the time would not qualify according to modern standards. They didn't have degrees in engineering. They started off as land surveyors . . . and worked into it."

The Act required graduation from a recognized university engineering program plus a minimum of two years of acceptable experience. (Today the requirement, for both engineers and geoscientists, is four years of acceptable experience.)

"Under the Act, equivalence to university graduation in engineering would be determined by a Board of Examiners at the University of Saskatchewan, based upon written examinations," wrote Taras Cheberiak in 1990, when he was general manager and registrar. "While many engineers of that time were good practical men, doing acceptable work in their own particular fields, it was clear that passing comprehensive examinations at the university level was not practical for them. Their opposition to the Act was overcome by a 'grandfathering' provision which conferred the rights of membership if they were, in fact, practicing engineering as defined in the proposed legislation at the time the Act came into force." (Cheberiak.)

Many of those "grandfathered" engineers went on to be very active in the early years of the Association.

Birth of the Association

The first meeting of the provisional council of APES, chaired by C.J. Mackenzie, Dean of Engineering at the University of Saskatchewan, was held on April 26, 1930, just five days before the passage of *The Engineering Profession Act* on May 1.

The members of the provisional council were issued the first nine registration numbers, with Number 1 going to Donald Alexander Roy McCannel, city engineer for the City of Regina. (Number 2 went to Mackenzie.)

At its first regular meeting in October, the provisional council adopted bylaws and a Code of Ethics, and appointed a nominating committee. The first annual meeting and election of officers took place on February 20, 1931. Mackenzie was elected president. By the end of 1931, the fledgling Association had 73 paid-up members.

One of the early concerns of engineers in Saskatchewan was that there suddenly seemed to be two organizations whose interests and activities overlapped: the Saskatchewan branch of the EIC and the new Association. That problem was solved with a joint agreement, signed in October 1938, which included an automatic membership arrangement (members of APES were automatically members of the Saskatchewan branch of EIC) and provision for common officers: Association councillors were also executive members of the Saskatchewan branch.

However, the primary concern of the Association was ensuring that those practising professional engineering in Saskatchewan were properly qualified.

"The trigger man"

The person ultimately responsible for enforcing the Act is the registrar of the Association. Originally, "it wasn't a paid position, but they paid a small honorarium because there were only 400 people in the organization at the time," said Bing-Wo. "I was a young fellow just fresh out of university with a degree in engineering, so I was keen on doing something for the profession and getting involved in its activities."

Bing-Wo's predecessors, like himself, were fully employed in some other capacity. (Bing-Wo was a design engineer with the Prairie Farm Rehabilitation Administration.) "For example, one was a superintendent of street railways," he said. "Another . . . worked for the City of Regina as an engineer."

But the Association found that it couldn't properly enforce the Act with part-time, essentially volunteer registrars, so in 1967 it hired Bing-Wo as its first full-time registrar and general manager.

Objects of Association

(from the 1930 bylaws)

To promote the public welfare by restricting the practice of professional engineering to persons who are or who may hereafter become properly qualified.

To promote the professional interests of practicing engineers.

To encourage all parties desirous of entering the Profession of Engineering to become fully qualified and to lend such help as from time to time, and in whatever manner may seem best to attain these ends.

Attendees at the Engineering Institute Convention held at the University of Saskatchewan August 8-10, 1918, pose for the camera. Proposed legislation in Manitoba, Saskatchewan and Alberta regarding the professional status of engineers was a hot topic of discussion among these delegates. *(University of Saskatchewan Archives A-2622.)*

Above: Representatives of Western branches of the Engineering Institute of Canada meet in Winnipeg en route to a national meeting in Montreal to discuss legislation establishing a body of professional engineers. H.R. MacKenzie, third from the right, was the Saskatchewan delegate. *(APEGS archival photo.)*

Right page (left): Hon. John E.N. Wiebe, Lieutenant Governor of Saskatchewan, signs the proclamation of *The Engineering and Geoscience Professions Act* on March 7, 1997, in Government House in Regina. Looking on (left to right) are Henry Feldkamp, then president of APEGS, Captain Walter Martin, Fran Haidl of the Geoscience Professional Registration Committee, and Inspector Cindy Villeneuve. *(Photo by Camera One.)*

Right page (right): APEGS President Bryan Schreiner poses with award recipients at the 1999 Annual Meeting. Left to right: Kevin Ansdell, Schreiner, Lorne Ritenburg, Ted Easton, Jim McCartan. *(APEGS archival photo.)*

It made a difference. "The legislation was always in place, but . . . they didn't have the full-time paid office help to enforce the Act, pursue it," Bing-Wo said. "When I came on board, one of the things I had to do was justify my paycheque, so I in fact started, shall we say, 'getting nasty' with people who were working in the province who weren't registered.

"I had the force of law behind me. If we could prove that they were in fact practising engineering, then of course you could bring a charge of illegal practice against them. When word got out that we were really going to enforce it, I think if you look at the membership figures, they started to grow. I don't want to take any credit for it . . . Of course I worked under the council of the Association and discipline committees and so on, but I was the trigger man."

It wasn't just individuals who had to be licensed, Bing-Wo said. "During the enforcement period, I think during the early days when I was active, people didn't take it too seriously, coming in to practise . . . Some client in Saskatchewan would hire some big firm from the States or something like that, and they thought, 'Well, we're the ABC Consulting Engineering Company, we'll go into Saskatchewan, we won't bother with (licensing).' But we tightened that right up."

Bing-Wo remembers people from one big US firm arriving at the border with "a big roll of plans. And the customs people stopped them. They said, 'Have you got a licence to practise in Saskatchewan?' They said, 'No, no I haven't.' 'Well, you can't bring that into Canada, if you're not licensed to practise.'

"I remember, they phoned me at home, this big firm. Of course, you've got to make them sweat a little bit . . . We finally got that cleared. It involved millions of dollars of work.

"You know, a doctor can't wander in here and (practise). Why should an engineer be able to do it?"

The prime directive

Which brings us to the defining aspect of being a professional engineer or geoscientist: his or her first responsibility is to the public.

Cheberiak put it this way: "While a professional engineer is obligated to serve the client or employer with diligence, there should be an awareness that the interests of society transcend those of individual persons or corporations. The engineer, though not always conscious of it, is dealing with public interest at every turn. Every problem ultimately comes down to three questions: 'Will it work?' 'Is it cost effective?' 'Is it safe?' The first two questions concern principally the client or employer, but the latter involves a universal and unceasing public obligation which no employer can ask the engineer to set aside." (Cheberiak.)

"The role of the Association is to protect the public from those who are not qualified to practise," said Bing-Wo. "It's the name of the game."

Members of the APEGS 1998-99 Council. Left to right, back row: Rick Forbes, Art Opseth, Brian Stronach, Clarence Reed, Doug Kozusko; second row, Anne Stephenson, Sandra Frattinger, Malcolm Reeves, Barry Fuchs, Sherri Praski, Fran Haidl; front row, Dan Hogan, Peter Miller, Bryan Schreiner, Margaret Kuzyk (Miller). Missing from photo: Jon Gillies. *(APEGS archival photo.)*

Stephenson, a registered nurse, was the first public appointee on the APEGS Council following proclamation of *The Engineering and Geoscience Professions Act*.

Geoscientists join the Association

Today, the Association also has the responsibility of ensuring that those practising professional geoscience are properly qualified. The Association of Professional Engineers of Saskatchewan (APES) became the Association of Professional Engineers and Geoscientists of Saskatchewan (APEGS) with the proclamation on March 7, 1997, of what is now called *The Engineering and Geoscience Professions Act*. The first geoscientists to be registered, at the proclamation ceremony, were Gary D. Delaney, Christopher F. Gilboy, Frances M. Haidl and Philip L. Reeves. Gilboy and Reeves were already registered as professional engineers.

There had always been some geoscientists in the association, because one section of *The Engineering Profession Act* enabled APES to register individuals with an honours degree in "a science related to engineering." Early on, physicists and chemists were registered under this provision, but later it was only used to register geoscientists. It was cumbersome, however. Geoscience applicants almost always ended up being assigned examinations in "complementary" subjects such as engineering economics, environmental issues, management and industrial relations. The amended Act and bylaws provided more appropriate criteria for registering geoscientists and a more appropriate designation to be conferred on them. It also brought geologists, who often work closely with engineers in the mining and oil industries (among others) under the provisions of the Act.

Gary Delaney, Director and Chief Geologist of Saskatchewan Industry and Resources' Northern Geological Survey, recalled that a few years before geoscientists joined the Association it became obvious that "professional registration was going to become *de rigueur* for geoscientists across the country, and we could either be proactive and become engaged in the process and help write the terms of reference in our engagement, or sit back and wait for government to say, 'You guys have got to get in there.'" So a group of geoscientists formed an ad hoc committee, with Delaney as chair, to walk through the process with the engineers.

The biggest challenge, Delaney said, was reassuring geoscientists they would be taken into the process as full and equal members. At that time, he said, geoscientists could get professional registration in Newfoundland, Alberta and British Columbia, but because the credit-hour system in use at the University of Regina was different from that in Alberta, "a number of our geoscientists couldn't get accepted in Alberta. So there was immediately a distrust of the whole process." But, said Delaney, "As a credit to the engineers, right from the get-go . . . they engaged us in the process, and certainly did as much as they could to accommodate our concerns."

The process took years. "I think we went through probably a dozen revisions of the Act," Delaney said, but the process went smoothly: even the sometimes contentious process

PROCLAMATION CEREMONY AND RECEPTION FOR
The Engineering and Geoscience Professions Act
IN THE PRESENCE OF THE
Honourable John E.N. Wiebe
Lieutenant Governor of Saskatchewan
and Her Honour, Mrs. Wiebe
GOVERNMENT HOUSE, REGINA
FRIDAY, MARCH 7, 1997

Program
Proclamation Ceremony
MASTERS OF CEREMONIES - RAYMOND S. PENTLAND, P.ENG.
PIANIST - IRENE SOTROPA, A.R.C.T.
OPENING REMARKS - MASTER OF CEREMONIES
ENTRY OF THE PLATFORM PARTY
ENTRY OF THE VICE-REGAL PARTY
REMARKS BY - HENRY F. FELDKAMP, P.ENG., PRESIDENT
READING OF PROCLAMATION
FRANCES HAIDL
GEOSCIENCE PROFESSIONAL REGISTRATION COMMITTEE
SIGNING OF THE PROCLAMATION
BY HIS HONOUR, THE LIEUTENANT GOVERNOR OF SASKATCHEWAN
REMARKS BY HIS HONOUR
REMARKS BY THE HON. ANDY RENAUD,
MINISTER OF HIGHWAYS AND TRANSPORTATION
DEPARTURE OF THE VICE-REGAL PARTY
Meeting of the Council
Receiving Line and Tour of Government House
Reception

Above: The program for the proclamation ceremony for *The Engineering and Geoscience Professions Act*, showing the newly designed crest representing both professions. *(APEGS archival photo.)*

of "grandparenting," bringing practising geologists into the association. "We had a committee that was mostly geoscientists, with I think one or two engineering representatives on it. Their philosophy right from the get-go was to be inclusive rather than exclusive, because if you get everybody in your membership, then you can address the concerns and issues and try to move forward."

Joining with the engineers also made economic sense for geoscientists, he said. "For a group of maybe two to three hundred people to set up their own association and have an office and support we would have had dues in the thousands of dollars."

There are two geological societies in the province that existed before geoscientists integrated with APEGS: the Saskatchewan Geological Society, based in Regina, and the Saskatoon branch of the Canadian Institute of Mining and Metallurgy Geological Society. But they're learned societies, not professional societies, Delaney explained. "For instance, the geological society here brings in a speaker every two weeks. Sometimes they will hold field trips or short courses to upgrade their members' skills, but they're not concerned with the area of professional practice, what a professional person can and can't do and how they should behave and things like that."

Delaney said it made sense for engineers and geoscientists to combine into one professional association, as is common right across the country (in all provinces but Ontario, Quebec and Nova Scotia) because there is considerable overlap in the kind of work they do. "There are a number of engineering professions, like geological engineers, geophysical engineers, geochemical engineers, where there's a logical cross-over."

"I think it's been a pretty good fit," he said. "Our members are certainly engaged in APEGS, they're on committees right across the board, one or two geoscientists have been president . . . I would say our relationship has been really good. I think we've benefited from it and they've certainly helped out our membership.

STOCK PHOTO

Impersonating an engineer . . . badly

Bing-Wo recounted one case involving an "engineer" who wasn't.

"I got a phone call from a mining company. They had a chap they hired as an engineer. He couldn't design a simple beam, which a second-year engineering student should be able to do.

"So we carried out an investigation. We found out in that particular case he had photocopied a Master of Science degree from McGill University, obliterated the original owner's name and put his name in, and changed the graduation date to suit his age.

"When they did the investigation, they found that the dean of engineering and the registrar, those guys were all dead or . . . weren't there any more, because the authentic holder of the diploma graduated many years before this person.

"Eventually he disappeared. . . . I understand he ended up in Alberta and they caught him there. He was doing the same thing. I think he went to jail for six months or something like that."

Above: Outgoing APEGS President Henry Feldkamp presents the chain of office to his successor, Margaret Miller, May 1997. *(APEGS archival photo.)*

"I think if all the associations across the country had the kind of support of management that this group has had, I think it would work out really well. Good leadership!"

"... there should be an awareness that the interests of society transcend those of individual persons or corporations."

A safe and prosperous future

Today, several thousand professional engineers and geoscientists are registered to practise in Saskatchewan. They work for Crown corporations and government departments, manufacturers and construction companies, oil companies and mining firms. Without them and their predecessors, there would be no roads, no buildings, no industry, no mines, no modern farms, no clean water, no electric light, no phones, no television . . . and no province, or at least not much of one.

From the vantage point of the province's past, we live in the future. That it is, indeed, a safe and prosperous one, is a tribute to the many engineers and geoscientists whose work is featured in this book—and also the many whose work is not.

A fully detailed history of engineering and geoscience in Saskatchewan would fill a thousand volumes.

But this one, at least, is a start.

CHAPTER 11 SOURCES

Websites

Association of Professional Engineers and Geoscientists of Saskatchewan. (April 17, 2006.) < http://www.apegs.sk.ca/>

Miscellaneous sources

Bing-Wo, Reginald. "The Engineering Profession in Saskatchewan." Unpublished manuscript provided to *The Engineering Journal*, April 25, 1976.

"Brief History of Professional Engineers of Saskatchewan." Undated, anonymous manuscript from APEGS archives.

Brief to the Government of the Province of Saskatchewan, presented by APES, March 1988.

Cheberiak, Taras. "Professional Engineering in Saskatchewan." Unpublished manuscript. March 1990.

"The Engineering Institute of Canada: 1887-1962." Document sent to branch chairmen of The Engineering Institute of Canada to assist with branch efforts to commemorate the 75th anniversary of the EIC. May 1962.

Legget, Robert F. "Engineering in Canada: Tracing our history pre-1887." *Canadian Consulting Engineer.* September/October 1987, pp. 16-17.

Minutes of Meetings of Council and of The Association of Professional Engineers of Saskatchewan from the time of Incorporation, May 1, 1930, to December 31, 1940, including A list of Certificates of Registration. APEGS archives.

Young, Stewart. "The Saskatchewan Branch, E.I.C.: 1915-1962." Unpublished manuscript, 1962.

Young, Stewart. "The Engineering Profession in Saskatchewan." Unpublished, undated manuscript.

Personal interviews by the author

Bing-Wo, Reginald. Former registrar of APES. December 8, 2005.

Delaney, Gary. Director and Chief Geologist, Northern Geological Survey, Saskatchewan Industry and Resources. Former chair of ad hoc committee exploring integration of geoscientists with APES. April 10, 2006.

INDEX

References to photos or captions are indicated by italics. In most cases, when an indexed entry appears in the caption, the page number given is for the photo the caption refers to rather than the page on which the caption appears.